Mitigating Vulnerability to High and Volatile Oil Prices

Mitigating Vulnerability to High and Volatile Oil Prices

Power Sector Experience in Latin America and the Caribbean

Rigoberto Ariel Yépez-García
Julie Dana

THE WORLD BANK
Washington, D.C.

ESMAP
Energy Sector Management Assistance Program

Contents

Boxes

Figures

Map

Tables

Foreword

The past decade has witnessed an unprecedented rise in world oil prices and oil price volatility. Between 2002 and 2012, the spot price for West Texas Intermediate increased more than fivefold, and this upward price trend featured significant volatility. In 2008, when oil prices reached their peak, the standard deviation in daily oil price changes was nearly twice that observed six years earlier. Oil importing and exporting nations alike are adversely affected by the greater economic uncertainty and higher risk introduced by oil price volatility. Countries with a high proportion of oil in their primary energy supply are especially vulnerable to higher and more volatile prices.

In the case of Latin America and the Caribbean (LAC), Central America and the Caribbean are net importers of crude oil and oil products. In both subregions, oil provides more than 90 percent of primary energy needs—more than one-third higher than the average for the LAC region overall. From 2002 to 2010, Central America saw the value of fuel imports double and the ratio of its average fuel imports to GDP increase by 2 percentage points, primarily as the result of higher oil prices.

This report offers a conceptual and practical assessment of how countries that are net oil importers can better cope with higher and more volatile oil prices. The study centers on the effects of such trends on the power sector. Although the examples used are limited to Central America and the Caribbean, oil-importing countries and industries in developing regions worldwide can draw general lessons.

The report delineates short-, medium-, and longer-term strategies to mitigate the effects of higher and more volatile oil prices in the energy sector. These range from financial instruments to lessen the impact of price volatility to structural measures—a more diversified power system, better energy efficiency in electricity production and use, and regional integration—to reduce the need for oil-based generation. For Central America alone, the estimated annual savings from regional electricity integration represent a reduction of about 8 percent in the oil-fired share of these countries' energy matrix. Such countries as Honduras, Nicaragua, and Jamaica would see the largest reductions in oil consumption by taking advantage of energy-efficiency strategies. Supply- and demand-side efficiency gains would lead to savings of up to 1 percent of GDP for Honduras, and nearly 1.5 percent of GDP for Nicaragua and Jamaica.

The aggregate effect of implementing these complementary strategies could significantly mitigate vulnerability to higher and more volatile oil prices. In Central America and the Caribbean, the total savings would equal about 35 percent of oil consumption for power generation. In terms of annual fuel purchases, it would amount to about 29 million and 11 million barrels of diesel and heavy fuel oil, respectively, representing about US$5 billion, based on the 2011 average price for these fuels. Some countries would witness significant reductions in their current account deficit—up to 5 percent of GDP—by implementing these initiatives in a combined strategy.

<table>
<tr><td>Ede Ijjasz Vasquez</td><td>Phillip R. D. Anderson</td></tr>
<tr><td>Director</td><td>Senior Manager</td></tr>
<tr><td>Sustainable Development Unit</td><td>Financial Advisory and Banking</td></tr>
<tr><td>Latin America and the Caribbean</td><td>Department</td></tr>
<tr><td>Region</td><td>World Bank Treasury</td></tr>
</table>

Acknowledgments

This report was prepared by Rigoberto Ariel Yépez-García of the Sustainable Energy Unit, Latin America and the Caribbean Region, and Julie Dana of the Financial Advisory and Banking Department, World Bank Treasury. The report benefited from the contributions of Luis San Vicente Portes, Claudio Alatorre, Alan Poole, Christopher Gilbert, Yuri Alcocer, Jose Luis Aburto, Donald Hertzmark, and David Santley. A special note of thanks goes to Shern Frederick for his valuable contributions and assistance in the editorial and analytical work and Gianfranco Bertozzi, Ricardo Tejada, and Dolores Lopez-Larroy for their input. The team is grateful to the peer reviewers, Marcelino Madrigal, Jasmin Chakeri, Charles M. Feinstein, David Reinstein, and David Santley, for their valuable comments. The team is grateful to Philippe Benoit for his valuable comments and overall guidance and support in preparing the report, including detailed review of multiple drafts. The team acknowledges Ede Ijjasz Vasquez, Jordan Schwartz, Malcolm Cosgrove-Davies, and Gregor Wolf for their valuable guidance and comments. The team thanks Aziz Gokdemir and Patricia Katayama, World Bank Office of the Publisher, for their guidance in producing the report. Finally, a special note of thanks goes to Norma Adams, editor, who took on the complex task of rewriting and editing the report and preparing it for publication.

The financial and technical support by the Energy Sector Management Assistance Program (ESMAP) is gratefully acknowledged. ESMAP is a global knowledge and technical assistance program administered by the World Bank. It provides analytical and advisory services to low- and middle-income countries to increase their know-how and institutional capacity to achieve environmentally sustainable energy solutions for poverty reduction and economic growth. ESMAP is funded by Australia, Austria, Denmark, Finland, France, Germany, Iceland, Lithuania, the Netherlands, Norway, Sweden, and the United Kingdom, as well as the World Bank.

Abbreviations

ACS	Association of Caribbean States
CAC	Central America and the Caribbean
CACG	Central America, the Caribbean, and Guyana
CAPP	Central American Power Pool
CARICOM	Caribbean Community
CARILEC	Caribbean Electric Utility Service Corporation
CCGT	Combined-Cycle Gas Turbine
CEAC	Central American Electrification Council (*Consejo de Electrificación de América Central*)
CFE	Federal Electricity Commission (*Comisión Federal de Electricidad*), Mexico
CFL	Compact Fluorescent Lamp
CIER	Regional Energy Integration Commission (*Consejo de Electrificación de América Central*)
CL	Commercial Losses
CNE-DR	National Energy Commission-Dominican Republic (*Comisión Nacional de Energía-República Dominicana*)
CNEE	National Electricity Commission (*Comisión Nacional de Energía Eléctrica*), Guatemala
ECLAC	UN Economic Commission for Latin America and the Caribbean
EIA	U.S. Energy Information Administration

ENEE	National Electricity Corporation (*Empresa Nacional de Energía Eléctrica*), Honduras
ESMAP	Energy Sector Management Assistance Program
FEIP	Oil Revenues Stabilization Fund (*Fondo de Estabilización de los Ingresos Petroleros*), Mexico
GDP	Gross Domestic Product
GHG	Greenhouse Gas
HFO	Heavy Fuel Oil
IBRD	International Bank for Reconstruction and Development
IDA	International Development Association
IGCC	Integrated Gasification Combined Cycle
IMF	International Monetary Fund
INDE	National Electrification Institute (*Instituto Nacional de Electrificación*), Guatemala
ISDA	International Swaps and Derivatives Association
LAC	Latin America and the Caribbean
LNG	Liquefied Natural Gas
LPG	Liquefied Petroleum Gas
MEM	Ministry of Energy and Mines (Peru)
MER	Regional Electricity Market (*Mercado Eléctrico Regional*), Central America
OLADE	Latin American Energy Organization (*Organización Latino Americano de Energía*)
OTC	Over the Counter
PIEM	Mesoamerican Energy Integration Program
PPP	Purchasing Power Parity
PV	Photovoltaic
RFP	Request for Proposals
RPP	Reference Price
SICA	Central America Integration System (*Sistema de la Integración Centroamericana*)
SIEPAC	Central American Electrical Interconnection System (*Sistema de Interconexión Eléctrica de los Países de América Central*)
TL	Technical Losses
TSF	Tariff Stabilization Fund
UN	United Nations
WASP	Wien Automatic System Planning
WTI	West Texas Intermediate

Units of Measure

bbl	U.S. barrel of oil (blue barrel) = 0.15898 m^3
boe	Barrel of oil equivalent
Btu	British thermal unit
GW	Gigawatt
GWh	Gigawatt hour
km	Kilometer
kV	Kilovolt
kW	Kilowatt
kWh	Kilowatt hour
kW$_p$	Kilowatt peak
MMBtu	Million British thermal units
MMSCFD	Million standard cubic feet
m/s	Meters per second
MW	Megawatt
TWh	Terawatt hour

Executive Summary

Countries heavily dependent on imported oil to power a significant portion of their electricity generation are especially vulnerable to high and volatile oil prices. In net oil-importing countries worldwide, high and volatile oil prices ripple through the power sector to numerous segments of the economy. As prices move up and down, so does the cost of electricity production, which has far-reaching effects on the economy, fiscal and trade balances, businesses, and household living standards.

High and volatile oil prices affect economies at both a macro and micro level. The major direct effects at the macro level are a deteriorating trade balance, through a higher import bill, reflecting a worsening in terms of trade; and a weakening fiscal balance due to greater government transfers and subsidies to insulate movements in international energy markets. At the micro level, investment uncertainty results from the higher risk of engaging in new projects and associated development and sunk costs, which, in turn, affects policy decisions and economic growth.

The major indirect effects are headline inflation, which may feed into core inflation through rising inflation expectations that trigger wage spirals; a loss of consumer confidence and purchasing power, due to greater economic uncertainty and higher inflation, which may reduce household discretionary spending and thus affect a major component of the economy; loss of competitiveness from higher power generation and transport costs, leading to decreased international competitiveness; and institutional weakening, as firms and households pressure the government to bypass market mechanisms, which, in turn, affects the credibility and functioning of the regulatory environment.

This study responds to the needs of policy makers and energy planners in oil-importing countries to better manage exposure to oil price risk. The study's objective is threefold. First, it analyzes the economic effects of higher and volatile prices on oil-importing countries, with emphasis on the power sector, using examples from Latin America and the Caribbean (LAC). Second, it proposes a menu of complementary options that can be applied over multiple time frames. Several structural measures are designed to reduce oil generation and consumption, while a range of financial instruments are suggested for managing price risk in the short term. Finally, it attempts to quantify some of the macroeconomic and microeconomic benefits that could accrue from implementing such options.

Oil Price Evolution and Risk Exposure

World oil prices have risen significantly over the past decade, and supply shocks have become increasingly common. Between 2002 and 2008, the spot price for West Texas Intermediate (WTI) increased sevenfold (from US$20 to 140 per barrel). This upward price trend also featured significant volatility. After peaking at $145 per barrel in July 2008, the WTI spot price fell sharply, bottoming at about $30 per barrel by year-end. By early 2011, prices had climbed back to $120 per barrel, mainly as a result of political unrest in supply countries of the Middle East and North Africa, along with technical problems in production; by mid-year, prices remained above $100 per barrel. The upward price trend may persist if limited spare capacity is required to serve additional demand from both developed and developing countries.

Various economic indicators can be used to measure a country's vulnerability to the economic effects of high and volatile oil prices. These include a greater share of oil imports as a percentage of gross domestic product (GDP), a high proportion of oil in the primary energy supply, and rising oil imports and expenditure over time. In the case of LAC, the region overall is a net exporter of crude oil and oil products; however, all countries in Central America and the Caribbean are net importers of these products (figure ES.1). In 2006, oil imports accounted for

Figure ES.1   Oil Trade Balance as a Percentage of GDP

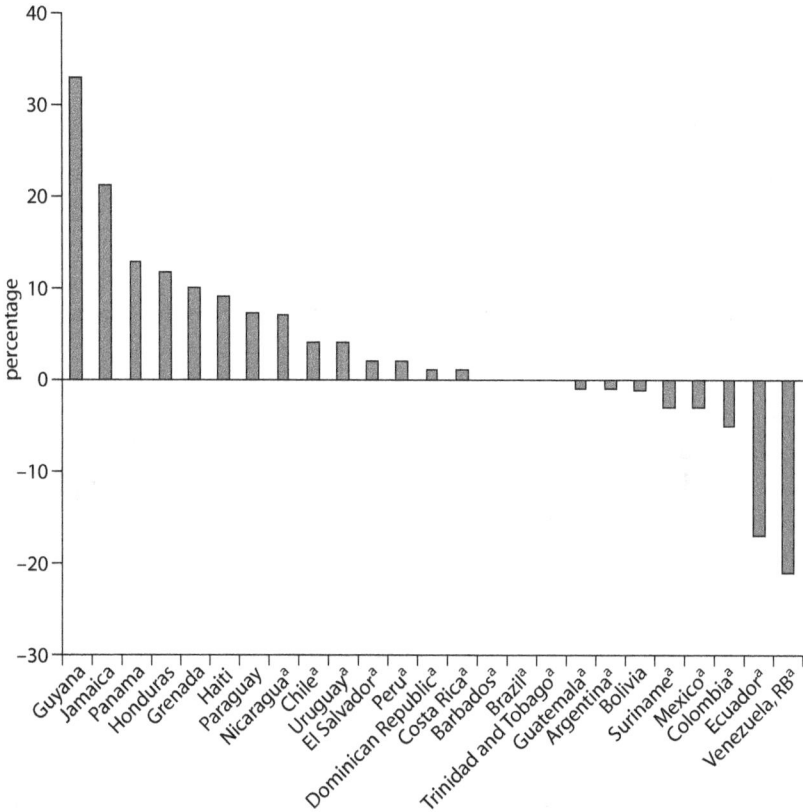

Sources: OLADE; EIA.
Note: a. Oil only, 2009 (OLADE); other observations, oil and oil products, 2006 (EIA).

8 percent and 11 percent of GDP, respectively. In both subregions, oil provides more than nine-tenths of primary energy needs—more than one-third higher than the average for the LAC region. In Central America in 2006–08, the value of oil imports rose by four-fifths and oil expenditure rose by 2.2 percent of GDP, primarily as the result of oil price increases.

Whether a country is a net oil exporter or importer often determines the direction and magnitude of the macroeconomic effects from higher oil prices. The World Bank estimates that, for the LAC region, a 16 percent increase annually in oil prices over a five-year period would increase growth in oil-exporting countries by 0.14 percentage points per year, compared to a loss of 0.10 percentage points per year for oil-importing countries. The greatest losses would occur in Central America and the Caribbean, at 0.09 and 0.12 percentage points per year, respectively.

Who Bears the Risk Burden?

Increasing subsidies during periods of high and volatile oil prices may lead to a deterioration of the fiscal balance. Government action during such periods often carries political risk, but general subsidies—especially those representing a significant share of government outlay—may lead to institutional weakening and budgetary stresses if not offset by expenditure cuts in other areas or higher taxes. If the government manages to maintain fiscal balance, the larger share of subsidies in government expenditure means less capacity for capital investment, as well as social and other programs. Broadly speaking, if subsidies are used to eliminate the impact of price volatility, government bears the risk burden. Conversely, if price volatility is passed through to customers, they bear the burden. Most countries share this risk between the two.

In a controlled energy-pricing environment with fixed consumer prices, the utility tends to absorb variations in price inputs. If state-owned, the utility passes the losses on to government, with notable effects on the fiscal balance. Conversely, in a free-market pricing environment, with a full pass-through mechanism, price shocks are passed on to consumer households and businesses. How the utility company is affected depends on the degree of demand elasticity and vertical integration. If demand is more inelastic in the short term, price changes can result in more revenue for the utility. The more vertically integrated the utility, the less intermediation cost it must absorb.

Most countries exhibit varying degrees of risk-sharing between consumers, the utility, and government. For example, the government might cap the electricity or fuel price for final consumers, making the private utility company bear the cost of price increases. But this situation may not be sustainable, as the company may eventually face bankruptcy. For utilities with both private- and public-sector interests, the government may have to bail out the utility or otherwise risk power-supply shortages.

In net oil-importing countries, managing oil price dynamics is a major challenge for the power sector. Power-sector policy decisions on managing oil price dynamics can have far-reaching economic effects. The complex interactions between a country's generation supply mix, electricity market structure and pricing policies, and utility ownership have budgetary and regulatory implications that can affect energy-sector planning and the ability to implement market-based solutions. In the case of Central America and the Caribbean, power generation relies mainly on oil products (diesel fuel and heavy fuel oil). All other generation sources—hydropower, geothermal, and biomass—together account for about two-fifths of overall power generation. The subregions' electricity market mainly features vertically integrated monopolies, wholesale competition, and single-buyer market structures with competitive generation. In countries where power sectors are dominated by vertically integrated, state-owned electric utilities, subsidies play a more important role. In most Caribbean countries, the government has the majority share of utility ownership, while the private sector predominates in Central America. For most of the 20 countries analyzed in this study, consumers are shielded to varying degrees by tariffs with embedded generalized subsidies.

Another management challenge for the power sector is making long-term electricity generation decisions in the face of the uncertainty created by oil price volatility. The planning and building out of new power generation capacity takes many years to achieve, requiring a framework for observing the effects of oil price changes on technology selection. The heightened uncertainty resulting from price volatility can cause energy-sector planners to delay investments or make inappropriate, sometimes irreversible generation-equipment decisions that affect electricity costs well into the future.

Reducing Short-Term Price Uncertainty

One option for managing oil price volatility in the short term is using price risk management instruments. Such tools can reduce the uncertainty associated with commodity-price volatility, particularly its impact on national budgets in a given year. The aim of such approaches is to manage existing price exposure, which is generally a function of current structural conditions. Risk management, or hedging, instruments are designed to cope with volatility—price spikes or shifting prices with no unidirectional trend—which has a financial impact because the existing price exposure results from the purchase of the physical commodity. Hedging should not be confused with *speculation*; the latter term refers to the use of price risk management instruments for the purposes of profiting from short-, medium-, or long-term price movements, independent of a direct interest in use of the physical commodity.

The two main categories of price risk management instruments are (i) physical and (ii) financial. Physical instruments include strategic pricing and timing of physical purchases and sales (e.g., "back-to-back" trading), forward contracts, minimum/maximum price forward contracts, price-to-be-fixed contracts, and long-term contracts with fixed or floating prices. Financial instruments include exchange-traded futures and options, over-the-counter options and swaps, collars, commodity-linked bonds, trade finance arrangements, and other commodity derivatives.

Though well-established in the commercial sector, the use of price risk management instruments is not widespread in the public sector, particularly by sovereigns. Recent volatility in energy and food prices, however, has awakened the interest of many governments that are eager to learn more about how they can use these tools. A critical first step for any country considering a commodity hedging strategy is careful risk assessment and evaluation of alternative hedging strategies. Given the complex commercial relationships in the power sector, along with the interactions between public and private actors, detailed risk assessment is critical. This can include (i) a supply-chain risk assessment that defines the roles and responsibilities of each actor in the sector, describing how each is affected by price volatility and (ii) a financial risk assessment that quantifies the price exposure resulting from specific commercial transactions or policy interventions and decisions. Equally important is ensuring that non-price related financial risks are isolated, monitored, and managed independent of direct price exposure.

Power-sector actors considering price risk management should focus on establishing an institutional framework that adequately supports implementation of the strategy. Key steps in the overall process of establishing a commodity hedging strategy include documentation of reasons for selecting a specific hedging product; establishment of the roles and responsibilities of various actors and agencies; verification of adequate legal and regulatory infrastructure; establishment of procedures for selecting counterparties and brokers; and careful oversight, supervision, and reporting.

Reducing Oil Consumption over the Long Run

Price risk management instruments cannot substitute for basic structural measures designed to reduce oil consumption over the longer term. The key structural instruments considered in this analysis are (i) energy portfolio diversification from oil-fired power generation, (ii) investing in energy efficiency, and (iii) increased regional integration with countries endowed with more diversified supply. These instruments provide the potential to reduce exposure to high and volatile oil prices, albeit with important limitations.

Diversifying from Oil-Fired Power Generation

In developing regions worldwide, concerns about climate change are spurring the development of renewable energy sources, which utilize local resources and produce cleaner energy. Renewables can also optimize the energy-generation portfolio because their cost is not correlated with oil prices, which could constitute up to 90 percent of the operating costs of certain generation technologies (e.g., a combustion-turbine plant using distillates). The cost of electricity generated from non-oil conventional fuels, such as natural gas, is somewhat correlated with oil prices, but much less so today than previously. Taken together, these benefits can reduce overall volatility. This conclusion is supported by recent studies that borrow from the portfolio models of the finance literature to determine and quantify the value from the optimal energy-generation system. By diversifying the power generation matrix, countries become less vulnerable to oil prices and reduce the risk attributed to oil price volatility.

Today, oil-importing countries have a wide array of choices—both renewables and non-oil conventional energy—for diversifying their energy generation portfolios away from oil. The potential for non-hydro renewables to comprise a greater share of power generation is significant. Biomass, in the form of sugarcane bagasse, could offer immediate output gains as long as appropriate retrofitting is put in place. From a policy perspective, geothermal has a large potential to diversify the power system, though exploration costs remain a barrier to resource exploitation. Other non-hydro options include wind and solar energy. In addition, non-oil conventional thermal power, particularly natural gas (and coal to a lesser extent), could help to reduce oil dependency, given their low price correlation with oil.

Improving Energy Efficiency

Investing in energy efficiency of both production (supply side) and end use (demand side) is one of the most cost-effective ways to reduce the need for oil and oil-derived products. The benefits are greatest for those countries that depend the most on oil-fired generation. On the supply side, reducing technical losses contributes to improving overall system efficiency and conserving fuel; thus, it is considered an instrument that directly mitigates exposure to oil price volatility. On the demand side, reducing peak and non-peak use helps to reduce the generation capacity and transmission and distribution assets required to supply the system.

Supply-side technical losses can be reduced by modifying system characteristics and configurations. These losses can also be reduced by carefully choosing transformer technology, eliminating transformation levels, switching off transformers, improving low-power factors, and distributing generation.

Demand-side efficiency can be improved by adopting policies and programs that encourage efficient electricity consumption by end users. Measures that could be expanded in the 20 countries analyzed in this study include standards for widely-used industrial equipment and residential appliances; building codes; consumer education and demonstration programs; and energy management programs for industry, the buildings sector, and public utilities.

Promoting Regional Integration

Regional energy integration can also help countries to reduce their oil dependence by optimizing electricity supplies across the region, which improves efficiency and, owing to economies of scale, lowers generation costs. In addition, when the consumption profiles of participants are not perfectly correlated, the smoother load pattern that arises means less investment in reserve requirements. If these conditions are met, use of fossil fuels, along with countries' vulnerability to high and volatile oil prices, declines. Furthermore, from a market perspective, regional integration promotes competition, helping to realize the trade gains associated with specialization of the most efficient producers. Moreover, all such benefits imply a reduction in greenhouse gas (GHG) emissions.

Looking ahead, Central America might become a corridor for a more robust interconnection between Colombia and Mexico. In the near future, both Mexico and Colombia are expected to have spare capacity available for export to Central America. The Colombia-Panama interconnection could be key to consolidating use of the new Central American Electrical Interconnection System (SIEPAC) infrastructure. Once SIEPAC becomes more consolidated, the possibility of effectively developing some of the region's larger hydro potential, as well as geothermal and wind energy, would be much improved.

Although the potential for interconnection in the Caribbean is more limited—owing to the high cost of needed submarine cables and small market size, which reduces economic viability—electricity integration could significantly reduce dependence on oil-fired generation. Interconnections between two or more countries could be economically feasible, and these would take advantage of economies of scale and development of indigenous resources. The geothermal and natural-gas potential of some islands could serve as the basis for a more diversified power market that is less vulnerable to oil prices. The Dominican Republic and Haiti, in particular, could benefit from stronger integration on both the power and natural-gas fronts.

While the economic benefits of integrated markets are generally accepted, institutional obstacles often prevent their establishment. The most common problems are use of multiple technology standards; variations in regulatory regimes, legal frameworks, and pricing policies; and environmental concerns. Additional hurdles that can limit or delay market integration are conflicting perspectives on the sharing of investment costs and uncertainties about political decision-making. In the case of SIEPAC, deep institutional differences have slowed progress on the harmonization of regulations. Another confounding factor has been the chronic shortage of generating capacity within countries, leading to a decline in intra-regional trade.

What Can Be Done

The aggregate effect of implementing these structural measures would significantly reduce the impact of high and volatile oil prices by reducing the need for oil-based generation. In both Central America and the Caribbean, renewable energy sources, including both hydropower and such non-hydro sources as geothermal and biomass, have considerable potential to comprise a greater share of power generation. In addition, greater energy supply- and demand-side efficiencies can result in potential fuel savings. Furthermore, a more integrated regional electricity market can allow for fuel savings by diversifying the power mix and achieving economies of scale. Taken together, these measures can achieve significant gains in energy security, as well as reduce GHG emissions.

The time frames for implementing measures to manage oil price volatility vary. For example, diversifying the energy portfolio through a greater share of renewables is a long-term measure, while investing in energy efficiency can be implemented over the medium or longer term. Over the short and medium term, financial instruments can be used to reduce exposure to price volatility (table ES.1).

Table ES.1   Overlapping Time Frames for Implementing Alternatives to Manage Oil Price Volatility

Alternatives for managing oil price volatility	Short term	Medium term	Long term
	Financial and physical hedging instruments		
	Energy efficiency		
			Hydro and non-hydro renewable energy and electricity trading

Source: Authors.

How Much Can It Help?

Implementing these several structural measures in a combined strategy would mean significant savings for heavily oil-dependent countries. In the case of Central America and the Caribbean, the fuel savings from a 10 percent increase in the potential generation capacity of renewable energy could amount to 14.2 million and 5.6 million barrels of diesel and heavy fuel oil (HFO), respectively, representing a reduction of several points of GDP in the countries' current account. By investing more in energy efficiency, the savings in barrels of diesel and HFO could total 3.5 million and 1.5 million on the supply side and 9 million and 2.4 million on the demand side. Finally, the estimated annual savings from regional electricity integration in Central America alone would amount to a reduction of about 8 percent in the oil-fired share of the countries' energy matrix. The combined savings from implementing all three measures in both subregions would be equivalent to about 35 percent of their oil consumption for power generation.

This optimistic outlook is not without its challenges. Making such a structural transition would entail considerable upfront costs to utilities, firms, and households; thus, supportive policies and regulations for renewable energy and energy efficiency would be required. In the case of the LAC region, regulatory, contracting, and licensing processes would need to be reformed to allow countries to implement their plans. Enabling financial instruments that make these investments possible would be helpful. Pricing reforms and technology standards would be needed to ensure that resources are not wasted. In addition, an appropriate regulatory framework and institutional strengthening would be required to facilitate regional integration between countries with differing regulatory policies and power-sector institutions.

But the potential benefits from implementing these measures far outweigh the costs. Given the far-reaching, adverse effects of high and volatile oil prices on oil-importing economies, the potential savings from implementing the measures suggested in this report could offer substantial benefits at the macro and micro level, ranging from long-term financial viability of the national economy to a higher living standard for households.

CHAPTER 1

Introduction

The power sectors of the world's oil-importing countries, particularly those with a large share of oil and oil products in their energy fuel mix, are vulnerable to high and volatile oil prices. As prices rise and fall, so does the cost of electricity production, which, in turn, affects the economy, fiscal and trade balances, businesses, and people's living standards.

World oil prices have risen significantly over the past decade. Between 2002 and 2008, the spot price for West Texas Intermediate (WTI), one of several major indicators of crude oil prices, increased sevenfold (from US$20 to 140 per barrel). This upward price trend also featured significant volatility (Bacon and Kojima 2008).[1] After peaking at $145 per barrel in July 2008, the WTI spot price fell sharply, bottoming at about $30 per barrel by year-end (figure 1.1).

By early 2011, prices had climbed back to US$120 per barrel, mainly as a result of political unrest in supply countries of the Middle East and North Africa, along with technical problems in oil production; by mid-year, prices remained above $100 per barrel. There is little reason to believe that price volatility will abate, given that supply shocks are increasingly common. And the upward price trend may persist if limited spare capacity is required to serve additional demand from both developed and developing countries.

Figure 1.1   Evolution of WTI Oil Prices and Volatility, 2000–12

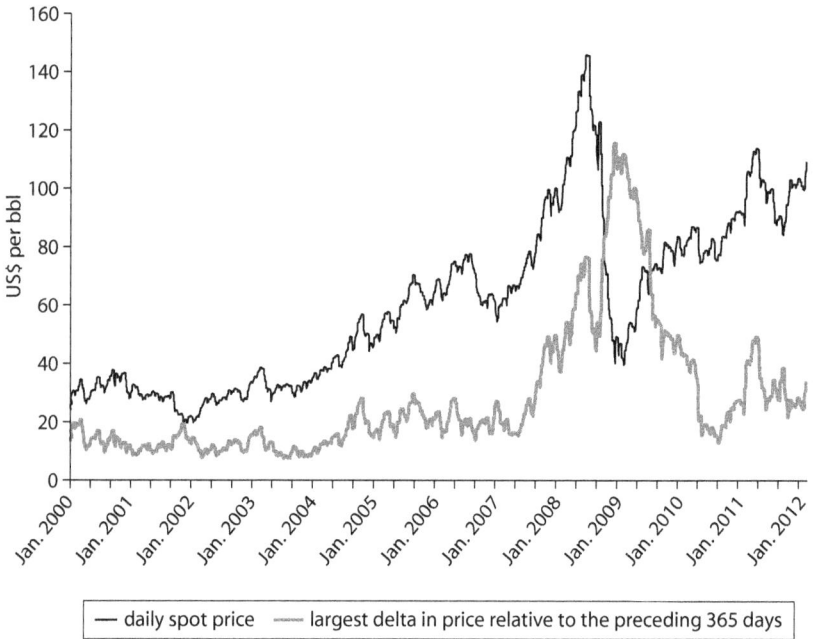

Source: Authors' estimation, based on EIA data.

Effects of High and Volatile Oil Prices

High oil prices variously affect economies, both directly and indirectly. At the macroeconomic level, they directly impact the aggregate economy. Government finances and balance of payments are affected, whether immediately or later on; inflationary effects and fiscal deficit may also be of concern. Indirectly, high oil prices may weaken the regulatory framework as governments implement nonmarket mechanisms, such as energy subsidies, to accommodate consumer demand for intervention. At the microeconomic level, utilities' energy planning ability and household purchasing power may be affected as higher oil prices are passed on to consumers. Firms' investment projects may become economically or financially unviable, while households may have less discretionary spending and experience an overall welfare loss.

Oil-exporting countries generally cheer a price increase, provided the adjustment in demand does not reduce overall oil revenues. For many oil-importing countries, however—particularly developing nations where oil comprises a large portion of the energy generation mix—high oil prices may trigger political unrest as the price of energy, which is critical for economic activity, is passed on to consumers. A second-order concern for oil-importing countries is price volatility. Volatile oil prices introduce greater risk and uncertainty in the profitability profile of investments using oil and oil-derived fuels. Firms may be forced to delay investment decisions, thus reducing capital formation and long-term growth. Despite their opposing preference on price-level direction, both oil-exporting and -importing nations dislike volatility because of the uncertainty created in the price level at which future sales and purchases are made.[2]

Study Background and Objective

This study builds on the results of two earlier World Bank–supported studies by Bacon and Kojima (2006, 2008). Based on their review of existing policy options to cope with higher oil prices in 38 developing countries worldwide, Bacon and Kojima (2006) concluded that eliminating traditional fuel subsidies that mainly benefit the wealthy would increase government revenue, remove pricing distortions, and reduce wasteful or nonessential energy use. They also suggested tackling demand management through fuel-saving measures as part of policies that provide multiple benefits (e.g., high-quality public transport). In countries where current prices contain some elements of subsidy, they suggested that governments persuade the public of the long-term cost-effectiveness of raising prices to market-clearing levels. To better help the poor, they recommended strengthening the databases used to more accurately identify low-income households and developing a delivery mechanism for income transfer and other types of well-targeted compensation.

In their examination of policy options for governments to better cope with oil price volatility, Bacon and Kojima (2008) noted that governments historically have made little use of hedging programs to manage fuel volatility. Focusing on the use of futures in managing price risk, they highlighted the basis risk and margin requirements that can make this form of risk management impractical for government entities.

Instruments Analyzed and Geographical Focus

While Bacon and Kojima (2008) indicated a limited role for hedging as a policy instrument, this study, drawing on recent experience with sovereign hedging, seeks to propose strategies that can address fiscal risk utilizing instruments, such as options contracts, which may be simpler for governments to use. Complementary to hedging instruments, the study considers several structural measures that policy makers and energy planners in oil-importing countries can utilize to better manage higher and more volatile oil prices over the medium and longer term.

The study is differentiated by its focus on Latin America and the Caribbean (LAC). In terms of oil price volatility, special focus is given to the challenges faced by island economies of the Caribbean, for whom hydrocarbon generation is nearly 100 percent of total generation by plant type. It explores the availability and use of diverse financial instruments to manage fuel price volatility in the shorter term, paying special attention to the needs of governments in the region.

Since region-specific proposals are an intended output of this study, it leverages previous research that has explored the feasibility of renewable energy sources to diversify generation for various locations in Central America and the Caribbean. Energy-efficiency improvements are examined at the level of power generation, transmission, and distribution as part of a comprehensive solution to reduce fuel requirements for a given electricity output. In addition, the potential for regional integration is considered to reduce end-user electricity prices by allowing countries with a comparative advantage in cost-effective generation to sell power to neighboring countries. For the purchasing countries, regional integration effectively diversifies their generation matrices; diversifying to a fuel with a different price volatility can also reduce the overall volatility of their fuel portfolio, making them less vulnerable overall to higher and more volatile oil prices.

Study Objective

This study's broad aim is threefold. First, it analyzes the economic effects of higher and volatile prices on oil-importing countries, with emphasis on the power sector, using examples from the LAC region. Second, it proposes a broad menu of alternatives—financial instruments to manage price risk, complemented by structural measures designed to reduce dependence on oil generation and consumption—that can be applied using a multi-horizon strategy. Finally, it attempts to quantify some of the macro and micro benefits that can accrue from implementing such alternatives.

Structure of This Report

This report is organized as follows. Chapter 2 differentiates the general effects and dynamics of high versus volatile oil prices in oil-importing countries and factors that determine how stakeholder groups are affected. Chapter 3 applies several economic indicators to determine a country's high oil dependence and thus vulnerability to high and volatile oil prices, using the LAC region as the focus of the analysis. Chapter 4 examines this vulnerability in the context of the region's power sector, including the market structure, utility ownership, and pricing policies. Chapter 5 details the price risk management (hedging) instruments that could be applied to manage oil price volatility over the shorter term. It offers a general overview of the benefits and drawbacks associated with the various instruments, and goes a step further by reviewing the process for assessing a country's commodity risk profile and making general recommendations about how to establish the institutional framework for commodity risk management.

The report's focus then shifts to the several longer-term structural measures to manage high and volatile oil prices by reducing oil consumption. These are diversification from oil in the energy generation mix, discussed in chapter 6; the role of energy efficiency in reducing electricity consumption, presented in chapter 7; and the relevance of regional energy integration to diversify energy sources, covered in chapter 8. Chapter 9 then attempts to quantify the potential benefits of these structural mitigation measures in terms of avoided oil consumption. Finally, chapter 10 offers concluding perspectives.

Notes

1. Volatility refers to frequent price fluctuations within a given time frame (e.g., daily deviations from the period average).
2. Thus, it is not surprising that the basics of proposed options to manage volatility specifically are transferrable between oil-exporting and oil-importing countries.

References

Bacon, R., and M. Kojima. 2006. *Coping with Higher Oil Prices*. Energy Sector Management Assistance Program (ESMAP) Report 323/06. Washington, DC: World Bank.

———. 2008. *Coping with Oil Price Volatility*. Energy Sector Management Assistance Program (ESMAP) Energy Security Special Report 005/08. Washington, DC: World Bank.

CHAPTER 2

Economic Effects of High and Volatile Oil Prices

The impact of high and volatile oil prices has been studied extensively. Such market changes impose a series of strains on the economies of oil-importing countries, affecting such diverse aspects as institutional strength, balance of payments, household spending, and social policy. The magnitude of these effects depends on the extent to which a country depends on energy imports and the diversification of its energy system. Sustained energy price volatility is likely to lead to less stable economic activity, which, in turn, can reduce investment and increase a country's perceived risk in international capital markets.

This chapter differentiates the specific effects of high and volatile oil prices on oil-importing countries, how these effects are interlinked, and their implications for economic growth and development. It then describes the major actors affected and factors that determine the risk burden of each group.

Effects of High Oil Prices

The economies of oil-importing countries are adversely affected by high oil prices at both the macroeconomic and microeconomic levels. At the macroeconomic level, indicators may include a deteriorating trade balance, inflation, and a bulging fiscal deficit; while the micro level may witness a reduction in real wages due to higher inflation, a greater portion of household income diverted to meeting higher fuel costs, and diminishing household savings and consumer confidence. The subsections that follow detail the specific effects and dynamics of high oil prices.

Trade Balance

High oil prices directly affect the trade balance of countries that must import oil for domestic consumption. In a net oil-importing country, the value of the country's exports decreases relative to that of its imports. This means that the country must export a larger quantity of goods to cover the amount of oil imported, all other factors being equal; otherwise, it must borrow from abroad or deplete foreign exchange reserves, which can develop into a balance-of-payments problem, putting pressure on the value of its currency.

These concerns are magnified in emerging economies that have high debt levels, large trade deficits, or difficulty tapping into capital markets. Such countries adjust by spending less, which negatively affects real economic activity. The IMF (2000) estimates that these effects would be larger in heavily indebted poor countries. Such countries run large trade deficits for lack of economic diversification, which, in turn, increase their dependence on consumption and capital goods from abroad. In addition, they are often on the margins of international capital markets.

Inflation

The persistence of high oil prices can feed directly into headline inflation via the pass-through of higher energy prices (e.g., oil products and electric power) to consumers or indirectly into core inflation, given that energy is a significant cost component in the production of goods and services (Barsky and Kilian 2004; Cavallo 2008).

The headline effect involves both inflation and rising expectations of inflation. To anchor inflation and preserve their credibility, the central banks must tighten interest rates at a time of weak demand, which increases the risk of recession. Batini and Tereanu (2009) explore a series of policy rules to apply under this type of scenario and the trade-off between credibility and stability under temporary oil price shocks. They conclude that inflation increases most in countries that take a softer stance on inflation targets, which, in turn, requires more stringent measures to bring inflation back to target and restore the central banks' credibility.[1]

The extent of the inflationary effect depends on the degree of the pass-through of energy prices to consumers. The IMF (2006) estimates that, among oil-importing developing economies, this effect is greater in countries of Sub-Saharan Africa, followed by those in Asia and the Western Hemisphere. A more limited pass-through can lead to deterioration of the fiscal balance due to energy subsidies. When final users are guaranteed a fixed price, there is a one-to-one impact on the fiscal balance. This means that, in net oil-importing countries, the gap between international and domestic oil prices must be covered by the government or power utility (depending on the ownership structure); otherwise, fuel shortages would result.

Competitiveness
Persistently high oil prices erode a country's comparative advantage in energy-intensive sectors. This is especially so in countries with an undiversified energy mix. In net oil-importing countries, the effect is direct as costs rise. And even in those countries rich in energy resources, a higher opportunity cost of the raw hydrocarbons can lead to de-industrialization akin to Dutch disease. Moreover, with rising inflation, the overall result may be an appreciation in the real exchange rate, thus affecting the competitiveness of the export sector as a whole, unless accompanied by a proportional depreciation of the nominal exchange rate (Chen and Chen 2007).

Consumer Confidence
Rising power and oil prices affect household-consumption and business-production decisions. From a welfare standpoint, higher energy prices decrease a household's purchasing power. Since consumers have limited flexibility to reduce fuel expenditures in the short run, a larger share of their income must be allocated to transport, heating, and electricity; thereby reducing all other purchases. Less disposable income implies less savings, especially for credit-constrained households for whom consumption smoothing is limited. Thus, in addition to higher costs that firms face, higher energy prices often result in diminished relative demand of most goods in favor of energy expenditures.

Balance of Payments
Less disposable income at the micro level due to high oil prices affects macro-level consumption. A drop in aggregate demand of the overall economy can be exacerbated by decreased investment associated with

worsening of firms' prospects. Other aggregate effects involve less savings, which can lead to higher funding costs and a reduction in national savings, which, in turn, can worsen the current account for a given level of investment. In time, this can accelerate a balance-of-payments crisis; even if the deficit can be externally funded by borrowing from abroad, it can increase the costs of debt and indirectly impact the government's finances and cost of capital for the private sector.

Fiscal Balance

In oil-importing countries, the government often plays an important role in the energy sector, including the pricing of oil products and electricity. The need to limit the cost to final users through subsidies leads to lower margins and eventually losses as oil prices increase power-generation and fuel-production costs, regardless of the ownership structure (public or private). Such policies, which are difficult to reverse owing to their political costs, perpetuate fiscal imbalances. A set of indirect effects also arises from lower economic activity, which, in turn, reduces tax receipts and increases transfers.

Institutional and Regulatory Framework

High oil prices can indirectly weaken an oil-importing country's institutional and regulatory framework as a result of public pressure by households, utilities, or firms for governments to resort to price controls and other nonmarket intervention mechanisms. However, the efficient functioning of the energy sector's institutional and regulatory framework depends on the time-consistent application of directives. Nonmarket interventions can diminish the regulatory framework's functionality and credibility and, because of higher consumer costs, impede return to a market-based pricing mechanism, which would likely face political opposition.

Effects of Oil Price Volatility

Volatile oil prices introduce uncertainty in the macroeconomic environment, which can reduce current spending; this, in turn, feeds into lower aggregate income, thereby worsening the impact of the initial price shock. The literature has often cited oil price volatility as the main force that depresses aggregate demand, owing to the transfer of income from net oil-importing to net oil-exporting nations (Ferderer 1996).

The investment uncertainty due to oil price volatility directly affects energy-sector planning, which, like other large-scale infrastructure investments, requires a long-term perspective. Even if future energy demand could be anticipated, it takes many years to plan and build out new energy sources. Heightened uncertainty and perceived risk can cause firms to delay investment decisions until prices stabilize, which can reduce capital formation and long-term economic growth.[2] Uncertainty adds risk to investment decisions in infrastructure, increasing the possibility of planners making inappropriate and sometimes irreversible choices that could affect energy costs well into the future. Not investing in the most appropriate technologies to generate affordable and efficient power may imply diverting productive resources to compensate for weaknesses in infrastructure and potentially constraining the development process.

What Is the Risk Distribution?

The ways in which households, utility companies, and governments are affected by high and volatile oil prices depend on the energy sector's market structure and the financial relationship between public- and private-sector stakeholders, including energy pricing policies. Depending on the sector's ownership structure and price-setting mechanism, price shocks are either passed on to consumers or otherwise mitigated (table 2.1).

In a free-market, energy-pricing environment with a full pass-through mechanism, price changes in fuels and electricity move through the supply chain, affecting consumer households and firms. How the utility company is affected depends on the degree of demand elasticity and vertical integration. If demand is more inelastic in the short term, price changes can result in more revenue for the utility company. The more vertically integrated the utility, the less intermediation costs it must absorb. Conversely, in a controlled pricing environment with fixed consumer prices, the utility tends to absorb variations in price inputs; if state-owned, the utility passes the losses on to government, with a notable effect on the fiscal balance (figure 2.1).

Table 2.1   Pricing, Utility Ownership, and Stakeholder Relationships

Pricing mechanism	Subsidy use	Utility ownership	Final cost burden
Fixed cost	Yes	Public sector (state-owned utility)	Government
Full pass-through	No	Private sector (electricity company)	Consumers
Shared cost	Yes	Public/private sector	Government

Source: Authors.

Figure 2.1 Distribution Scenarios of Higher Energy Costs

Who pays the added cost?

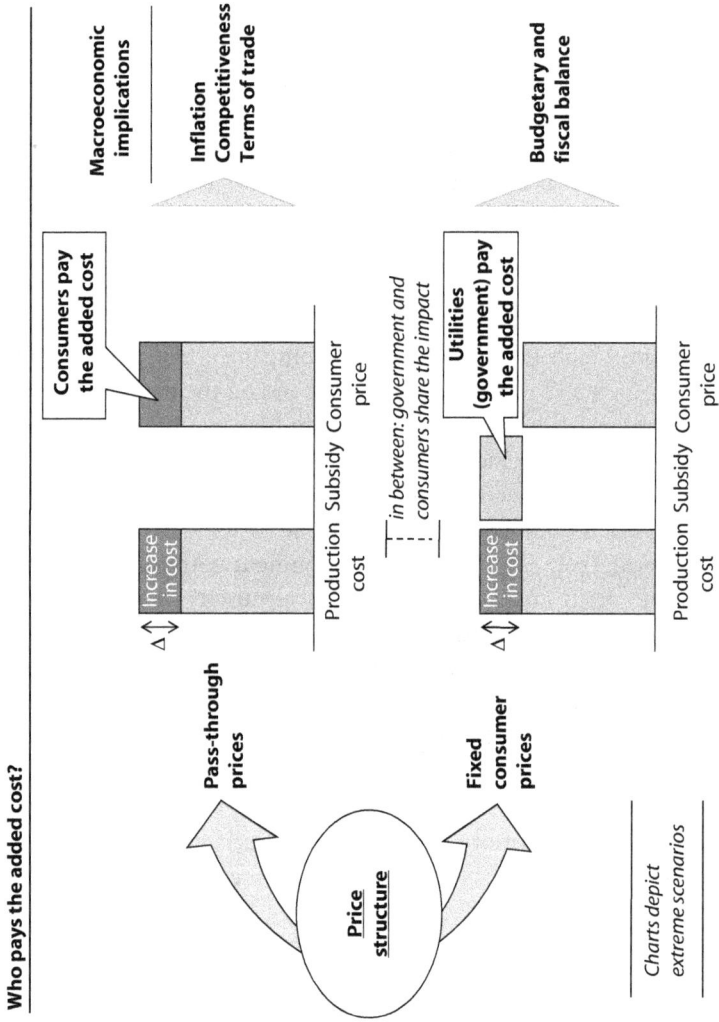

Source: Authors.

Most countries exhibit a mix of the above-described scenarios, with varying degrees of risk-sharing between consumers, the utility, and government. For example, the government might cap the electricity or fuel price for final consumers, making the private utility company bear the cost of price increases. But this situation may not be sustainable, as the company may eventually face bankruptcy. For utilities with both private- and public-sector interests, the government may have to bail out the utility or otherwise risk power-supply shortages.

Countries are affected by the distribution of volatility risk among stakeholder groups. They can be classified according to the group that bears the cost, which depends on the degree to which tariffs are subsidized. For example, consumers bear the cost of volatility risk if no subsidies are applied, while the risk to government grows as increasingly larger subsidies are applied. Most governments intervene with subsidies or other smoothing mechanisms, which are integrated with policies to

---

Box 2.1

Reducing the Cost Burden of Subsidies

A 2006 study by Bacon and Kojima found that subsidies are a common price-based policy instrument for coping with higher oil prices. A review of 38 developing countries worldwide showed that subsidy mechanisms may include direct subsidies to consumer groups, indirect subsidies through lowered taxes on petroleum products, and targeted income subsidies. Traditional fuel subsidies were found to have large leakage, resulting in low cost-effectiveness. Household surveys confirmed that the lowest income groups often receive the smallest share of the subsidy benefits.

The authors concluded that eliminating poorly targeted subsidies that mainly benefit the wealthy would increase government revenue, remove pricing distortions, and reduce wasteful or nonessential energy use. In countries where current prices contain some elements of subsidy, the authors recommended that government strategy persuade the public of the long-term cost-effectiveness of raising prices to market-clearing levels. To better address the needs of the poor, governments would be well advised to strengthen the database used to identify poor households and develop a delivery mechanism for income transfer and other types of compensation that better targets low-income households.

Source: Bacon and Kojima 2006.

set fuel or electricity prices. The increase in fuel prices in 2007–08, for example, led many governments to intervene, either directly or indirectly, in setting fuel or electricity prices (Kojima 2009). Such subsidies are financed through various mechanisms, including direct transfers from government budgets and cross-subsidies built into fuel prices and electricity rates.

When energy subsidies represent a significant share of government outlay, they increase the vulnerability of the government's finances to oil price volatility. Unless offset by expenditure cuts in other areas or higher taxes, such expenses can deteriorate the fiscal balance and increase public debt. If the fiscal balance is held relatively constant, a larger share of subsidy in government expenditure means less room for capital improvement, both within and outside the energy sector, and social expenditure. In addition, subsidies can contribute to exacerbating the impact of oil price volatility on the economy by encouraging more consumption (box 2.1).

Summary Remarks

This chapter has demonstrated the breadth and depth of the adverse effects that high and volatile oil prices can have on net oil-importing countries. High prices may lead to trade and fiscal imbalances, a crisis of consumer confidence and rising inflation, as well as a weakening of competition and the regulatory framework; while price volatility creates uncertainty in energy planning and investment, which affects economic growth. In light of the variation in the timing and duration of these problems—ranging from short-term hindrances to permanent changes in the macroeconomy—an effective solution calls for a multi-horizon strategy.

Furthermore, the degree of demand elasticity for electricity and vertical integration in the sector influences how utilities are affected by high and volatile prices. In a controlled pricing environment with fixed consumer prices, the utility might absorb variations in price inputs in the short run, but these are unsustainable in the longer term. If state-owned, it passes the losses on to government, with an important effect on the fiscal balance. The next chapter considers some of the economic indicators that help to determine whether a country exhibits high oil dependence and thus greater risk exposure to the economic effects of high and volatile oil prices discussed in this chapter. The focus of the analysis is Latin America and the Caribbean.

Notes

1. The primary recommendations of Batini and Tereanu (2009) are that central bank transparency and a good communication strategy are necessary to maintain inflation expectations; from a practical standpoint, they find that early responses to shocks lead to small output changes over time.

2. Bacon and Kojima (2008a, 2008b) have analyzed the effect on such macroeconomic variables as an economy's oil-shock vulnerability (defined as the ratio of the value of net oil imports to GDP), terms of trade, a government's overall financial surplus relative to GDP, and the ratio of debt to GDP.

References

Bacon, R., and M. Kojima. 2006. *Coping with Higher Oil Prices*. Energy Sector Management Assistance Program (ESMAP) Report 323/06. Washington, DC: World Bank.

———. 2008a. *Oil Price Risks: Measuring the Vulnerability of Oil Importers*. Public Policy for the Private Sector, Note No. 320. Washington, DC: World Bank Group.

———. 2008b. *Vulnerability to Oil Price Increases: A Decomposition Analysis of 161 Countries*. Extractive Industries and Development Series #1. Washington, DC: World Bank Group.

Barsky, R., and L. Kilian. 2004. "Oil and the Macroeconomy Since the 1970s." *Journal of Economic Perspectives* 18(4): 115–34.

Batini, N., and E. Tereanu. 2009. "What Should Inflation Targeting Countries Do When Oil Prices Rise and Drop Fast?" IMF Working Paper WP/09/101, International Monetary Fund, Washington, DC.

Cavallo, M. 2008. *Oil Prices and Inflation*. FRBSF Economic Letter No. 2008-31. San Francisco, CA: Federal Reserve Bank of San Francisco.

Chen, S. S., and H. C. Chen. 2007. "Oil Prices and Real Exchange Rates." *Energy Economics* 29(3): 390–404.

Ferderer, P. 1996. "Oil Price Volatility and the Macroeconomy." *Journal of Macroeconomics* 18(1): 1–26.

IMF (International Monetary Fund). 2000. "Impact of Higher Oil Prices on the Global Economy." Research Department Staff Paper, International Monetary Fund, Washington, DC.

———. 2006. *Regional Economic Outlook: Sub-Saharan Africa*. World Economic and Financial Surveys. Washington, DC: International Monetary Fund.

Kojima, M. 2009. *Government Responses to Oil Price Volatility: Experience of 49 Developing Countries*. Extractive Industries for Development Series #10. Washington, DC: World Bank.

CHAPTER 3

Economic Indicators of Vulnerability: Analysis of Latin America and the Caribbean

A country's vulnerability to high and volatile oil prices is determined, in part, by the degree to which it depends on oil imports, the proportion of oil in its primary energy supply, and rising oil expenditures as a share of GDP. One of the most frequently cited indicators in the literature for measuring vulnerability is the ratio of the value of net oil imports to GDP. If oil prices and/or consumption rise, the value of net oil imports increases and the ratio rises, making countries more vulnerable.

In this chapter, we use the above-mentioned economic indicators to analyze countries in the Latin America and the Caribbean (LAC) region. By comparing the country values of these indicators with global, regional, and subregional ones, we aim to identify which countries are most vulnerable to high and volatile prices and thus the economic effects discussed in chapter 2.

Oil Trade Dynamics

Countries can be distinguished as oil and gas exporters, net importers, or balanced (i.e., having an oil-and-gas trade balance close to zero). In the LAC region, oil-and-gas exporting countries include Colombia, Ecuador, Mexico, Trinidad and Tobago, and República Bolivariana de

Venezuela. Balanced countries comprise Argentina, Bolivia, Brazil, and Peru; while net oil-importing countries include Chile, Paraguay, and Uruguay, as well as most countries in the subregions of the Caribbean and Central America.

Oil Imports as a Share of GDP

Countries with a greater share of oil imports as a percentage of GDP generally exhibit greater vulnerability to high and volatile oil prices. In the case of the LAC region, the Caribbean nations historically have exhibited the greatest reliance on oil, followed by countries in Central America. In 2006, oil imports accounted for an average of 11 percent and 8 percent of GDP for the Caribbean nations and Central America, respectively, compared to 3 percent in oil exports for the LAC region overall (figure 3.1); thus, these countries' case deserves special attention.

Another measure of a country's vulnerability to oil prices is the ratio of net oil imports to GDP. The LAC region overall is a net exporter of crude oil and oil products; yet all countries in both the Caribbean and Central America are net importers of these products (figure 3.2).[1]

Figure 3.1   Country and Subregional Comparisons of Oil Imports as Share of GDP

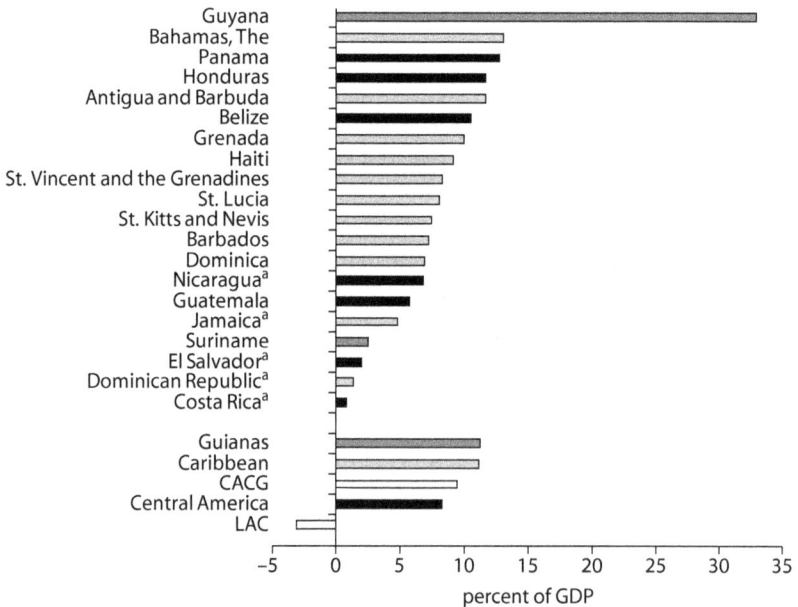

Sources: OLADE; EIA.

Note: a. Oil only, 2009 (OLADE); other observations, oil and oil products, 2006 (EIA).

Figure 3.2   Ratio of Net Oil Imports or Exports to GDP in LAC Countries

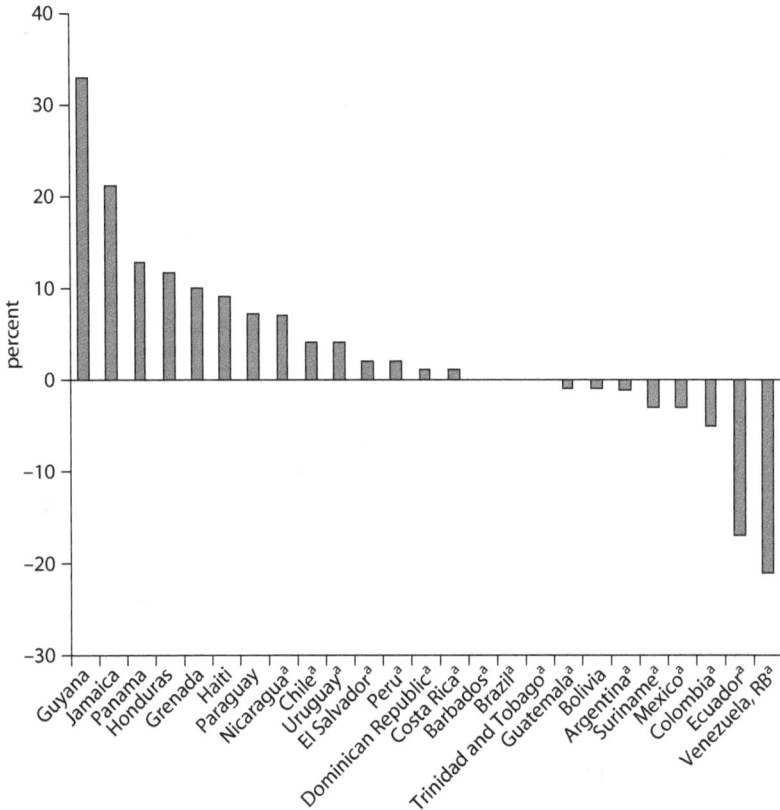

Sources: OLADE; EIA.
Note: a. Oil only, 2009 (OLADE); other observations, oil and oil products, 2006 (EIA).

Oil as a Share of Primary Energy Supply

In Central America and the Caribbean, oil supplies 51 percent of primary energy needs, compared to 42 percent for the LAC region, and 35 percent for the world overall (figure 3.3).

Four of the 20 countries examined in Central America and the Caribbean—Barbados, Belize, Guatemala, and Suriname—produce oil, but production is insufficient to meet domestic demand.[2] The other 16 countries depend entirely on imports to cover domestic demand for oil products; some have refineries that use imported crude oil to produce some fraction of this internal demand. In response to high oil price vulnerability in Central America and the Caribbean, Mexico and República

Figure 3.3   Comparisons of Primary Energy Supply by Source, 2008

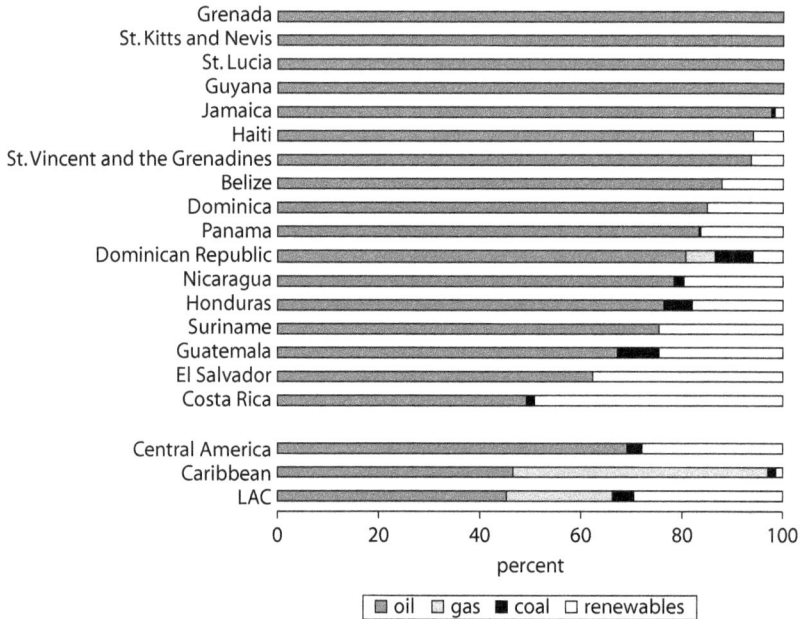

Source: Authors' calculations, based on EIA data.

Bolivariana de Venezuela have led regional efforts to supply the subregions oil and fuels at a preferential price (box 3.1).

Rising Oil Imports and Expenditure over Time

Beyond comparing the ratio of imported oil to GDP or oil-based energy supply at specific points in time, it is also important to view such vulnerability indicators over time. For example, over a two-year period (2006–08), oil imports rose by four-fifths in Central America, primarily as the result of price increases;[3] over that same period, oil imports in Costa Rica alone increased 124 percent and nearly doubled in Honduras and Panama (table 3.1).[4]

Rising oil expenditure as a share of GDP is another indicator of vulnerability to high and volatile oil prices. In Central America, the increase was 2.2 percent of GDP over a two-year period as a result of a large jump in price. The largest percentage changes occurred in Honduras and Nicaragua (table 3.2).

The year 2008 was unique in that oil prices had reached a level not seen since the early 1980s in inflation-adjusted terms.[5] Elevated levels in

Box 3.1

Regional Energy Cooperation for Socioeconomic Development

The Energy Cooperation Treaty of San José, signed by Mexico and República Bolivariana de Venezuela in 1980, is a regional effort to promote social and economic development in Central America and the Caribbean. Under the agreement, each country sells beneficiary countries 80,000 barrels of oil and refined oil products at preferential prices, with preferential financing terms for development projects. The beneficiary countries are Barbados, Belize, Costa Rica, the Dominican Republic, El Salvador, Guatemala, Haiti, Honduras, Jamaica, Nicaragua, and Panama. Under a joint declaration, the agreement was renewed by governments of the two countries in 2007. A similar scheme, the Caracas Energy Cooperation Agreement, was signed by República Bolivariana de Venezuela in 2000 to include more Caribbean countries and provide longer-term financial facilities for oil purchases.

Petrocaribe, an alliance between República Bolivariana de Venezuela and a number of Caribbean countries, allows participating countries to buy Venezuelan oil on preferential payment terms. The financing scheme consists in buying the oil at the reference price, but on favorable credit terms. Member countries include Antigua and Barbuda, The Bahamas, Belize, Cuba, Dominica, the Dominican Republic, Grenada, Guatemala, Guyana, Haiti, Jamaica, St. Kitts and Nevis, St. Lucia, St. Vincent and the Grenadines, and Suriname.

Source: Authors, with Petrocaribe data (www.petrocaribe.org).

oil expenditure as a share of GDP help to explain the responses of governments in the region. For example, when prices rose, Honduras decided to abandon an automatic end-user, price-adjustment mechanism; importers, who received inadequate compensation for fuel sales, became financially constrained and, in turn, reduced purchases, which eventually led to fuel shortages. In Panama, where the market price of fuel oil was higher than the price established in the tariff regime, the government spent US$13.4 million to cover the over-cost.

Oil Exporter versus Importer

Whether a country is a net oil exporter or importer often determines the direction and magnitude of the macroeconomic effects from higher oil prices. For the LAC region, the World Bank (2006) estimates that a

Table 3.1   Oil Imports in Central America, 2006–08

Country	Oil imports (US$, millions)			Change, 2006–08 (%)
	2006	2007	2008	
Costa Rica	1,250.0	1,452.0	2,800.0	124
El Salvador	1,000.0	1,288.0	1,680.0	68
Guatemala	1,876.5	2,418.4	2,973.1	58
Honduras	1,088.5	1,375.7	2,000.0	84
Nicaragua	689.7	836.5	1,133.0	64
Panama	1,065.0	1,231.4	2,026.0	90
All Central America	6,969.7	8,602.0	12,612.1	81

Source: ECLAC 2009.

Table 3.2   Oil Expenditure as Share of GDP in Central America, 2006–08

Country	Oil expenditure as share of GDP (%)			Change, 2006–07 (%)	Change, 2007–08 (%)
	2006	2007	2008		
Costa Rica	5.6	5.5	8.8	−0.1	3.3
El Salvador	6.4	7.4	7.7	1.0	0.3
Guatemala	6.1	6.6	7.8	0.5	1.2
Honduras	10.1	10.6	14.5	0.5	3.9
Nicaragua	13.0	14.7	18.6	1.7	3.9
Panama	5.2	6.2	9.3	1.0	3.1
Central America (average)	6.6	7.3	9.5	0.7	2.2

Sources: ECLAC 2009; data from central banks.

16 percent increase annually in oil prices over a five-year period would increase growth in oil-exporting countries by 0.14 percentage points, compared to a loss of 0.10 percentage points for oil-importing countries. The greatest losses would be experienced in the subregions of the Caribbean and Central America, at 0.12 and 0.09 percentage points, respectively (figure 3.4). For some countries, the macroeconomic effect is unclear; for example, an oil-exporting country may use windfall oil profits to subsidize fuel prices or reduce fiscal deficit; however, such cases are the exception (table 3.3).

As table 3.3 suggests, the pass-through of fuel prices into inflation is limited, as is the response of monetary policy; these phenomena can be explained, in part, by the perception of monetary authorities that oil price shocks are temporary. However, countries in Central America and the Caribbean, notably net oil importers, experience a deterioration of their fiscal balance because of larger subsidies or lower tax receipts due

Figure 3.4   Growth Effect Comparisons from Higher Oil Prices in Latin America and the Caribbean

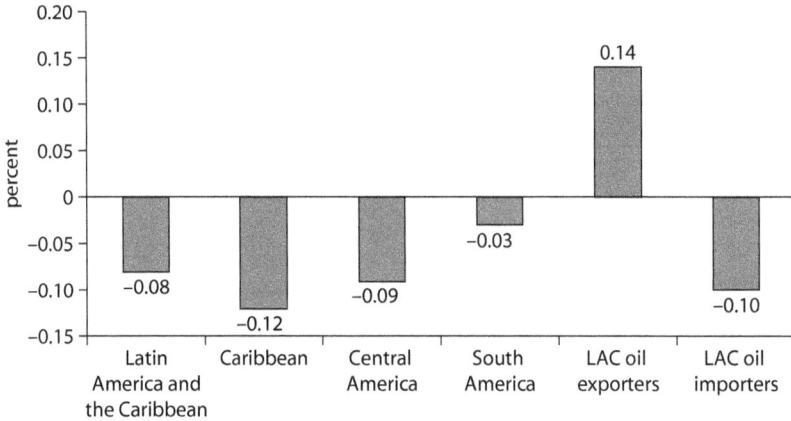

Source: World Bank 2006.

Table 3.3   Macroeconomic Effects of Higher Oil Prices in Selected Countries of Latin America and the Caribbean

Country	CPI pass-through	Fiscal balance	Interest rates	External accounts
Argentina	None	−	None	+
Brazil	Limited	+	None	+
Colombia	None	+	None	+
Dominican Republic	Significant	−	Significant	−
Ecuador	None	+	—	+
El Salvador	Limited	−	—	−
Guyana	Limited	−	Modest	−
Honduras	Significant	−	None	−
Mexico	None	+	None	+
Venezuela, RB	None	+	None	+

Source: World Bank 2006.
Note: CPI = consumer price index, — = not available.

to an economic slowdown. In addition, these countries' external and fiscal effects are symmetric, as higher oil prices translate into higher energy import bills.

Dynamics of Oil Price Hikes and Subsidies

Government inaction during periods of high and volatile oil prices may carry political consequences; yet increasing energy subsidies during such periods may lead to the institutional weakening and budgetary stresses discussed in chapter 2. If not offset by expenditure cuts in other areas or higher taxes, subsidies—especially those that represent a significant share of government outlay—can cause deterioration of the fiscal balance or otherwise risk increasing the public debt. If the government manages to maintain fiscal balance, the larger share of subsidies in government expenditure may mean less capacity for capital investment, as well as social and other programs. Increasing general subsidies during periods of high and volatile oil prices may exacerbate adverse economic effects. Higher subsidies would affect taxpayers and distort private-sector investment decisions, which, in turn, would affect the generation matrix, further exposing countries to high and volatile oil prices.

From a social welfare perspective, the application of well-targeted government subsides can help to alleviate poverty. Deciding between implementing a policy of subsidized prices or an alternative social policy aimed at granting equivalent transfers directly to a target population has important implications for businesses, consumers, taxpayers, poor households, and the government. No doubt, the best alternative is one that both reduces economic distortions and effectively supports the population group for whom the subsidy is intended.

In the case of Central America following the 2008 oil price hike, five of the six countries chose to introduce or provide additional subsidies. In El Salvador, for example, government expenditure for energy subsidies more than doubled. Costa Rica was an exception; despite the sharp price rise, it did not introduce subsidies (table 3.4).

Table 3.4   Change in Energy Subsidies in Central America, 2007–08

| Country | Energy subsidies (US$, millions) | | Change (%) |
	2007	2008	
Costa Rica	0	0	0
El Salvador	194.5	420.1	116
Guatemala	—	86.2	—
Honduras	118.4	218.6	85
Nicaragua	—	67.7	—
Panama	97.6	165.3	69

Sources: ECLAC 2009; data from country finance ministries.
Note: — = not available.

Summary Remarks

This chapter has illustrated some of the economic indicators that help to determine whether a country is highly oil-dependent and thus vulnerable to the economic effects of high and volatile oil prices. Analysis of experience in the LAC region demonstrates that vulnerability is concentrated mainly in the subregions of Central America and the Caribbean. The next chapter examines how this vulnerability is exhibited in the power sector.

Notes

1. Unlike its neighbors, Trinidad and Tobago fuels its power generation almost entirely from natural gas. As a fuel exporter, the country is also vulnerable to changes in oil prices, but in a unique way; owing to its special features, it is excluded from the set of 20 countries examined in Central America and the Caribbean (appendix A).

2. Suriname meets most of its internal demand with fuels produced in the local refinery from locally-produced crude oil, but has a deficit of 15 percent (measured in volume units). Barbados, Belize, and Guatemala import all of their oil products, a small portion of which is offset by crude-oil export revenues; Barbados imports 93 percent of total consumption, Guatemala 72 percent, and Belize 69 percent (EIA 2006). The degree to which these oil-producing economies are vulnerable to oil price changes depends on how the resource rent is shared between foreign companies, locally-owned ones, and governments; that is, how much of each additional US$1 per bbl in the international price of oil remains in the local economy or is transferred abroad, which, in turn, depends on the ownership of oil-producing facilities, as well as oil agreements and related fiscal rules.

3. Price increase is the more likely cause of the rise in import value since annual energy intensity in Costa Rica decreased slightly from 2006 to 2008 (from 8,600 to 8,200 Btu per year [2005 US$]).

4. Similarly, from 2006 to 2008, energy intensity decreased in Honduras (from 12,100 to 11,600 Btu) and Panama (from 13,700 to 11,500 Btu).

5. The monthly average oil price peaked in December 1979 at US$108.59 per barrel (January 2011 US$).

References

ECLAC (UN Economic Commission for Latin America and the Caribbean). 2009. *La Crisis de los Precios del Petróleo y su Impacto en los Países Centroamericanos.* Report No. LC/MEX/L.908, June 18. Mexico City: UN Economic Commission for Latin America and the Caribbean.

EIA (U.S. Energy Information Administration). 2006. Statistical Database. http://www.eia.gov.

World Bank. 2006. *Assessing the Impact of Higher Oil Prices in Latin America.* Economic Policy Sector. Washington, DC: World Bank.

CHAPTER 4

Managing Oil Price Dynamics in the Power Sector: Experience in Latin America and the Caribbean

In net oil-importing countries, high and volatile oil prices ripple through the power sector to numerous segments of the economy. Power-sector policy decisions on the best ways to manage price dynamics can have far-reaching economic effects. The complex interactions between such a country's electricity-generation mix, market structure, and utility ownership have budgetary and regulatory implications that affect energy-sector planning and the ability to implement market-based solutions.

Oil-Dependent Energy Mix and Cost of Power Generation

More than half of power generation in the subregions of Central America and the Caribbean is oil-based, at about 38 percent and 75 percent, respectively (figure 4.1).[1]

Such countries can experience cost increases of up to 3.5US¢ per kWh for every US$10 increase in the price of oil. Figure 4.2 suggests the average tariff increase that should accompany price increases to maintain power-system profitability in these countries.

As the impact of high and volatile oil prices ripples through the power sector, mitigating vulnerability presents a complex management challenge, as Guatemala's experience illustrates (box 4.1).

Figure 4.1   Electricity Generation Mix in Selected Countries and Regions, 2007

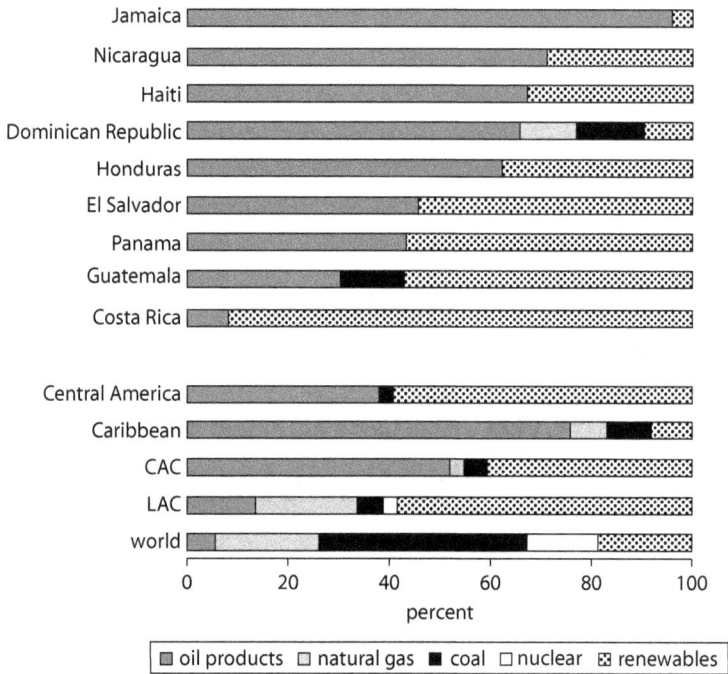

Source: Authors, based on OLADE and EIA data.
Note: In Guatemala, renewables could represent a somewhat smaller portion (about 36 percent) of the electricity generation mix.

Distribution of Costs and Risks

Electricity Market Structure

The structure of a country's electricity market and utility ownership affects how costs and risks are distributed across the electricity sector. In the case of Central America and the Caribbean, three models characterize the market structure: (i) vertically integrated monopoly, (ii) wholesale competition, and (iii) single buyer with competitive generation (table 4.1).

In a **vertically integrated monopoly**, generation, transmission, distribution, and systems operations are treated as an integrated whole. The national power systems are generally too small for competition to result in significant gains for consumers. This is the model used in 9 of the 11 Caribbean countries considered in this study. In the **wholesale competition** model, large commercial and industrial consumers compete to procure electricity directly from independent generators or through

Figure 4.2   Impact of Oil Price Changes on Power Generation Costs, 2006

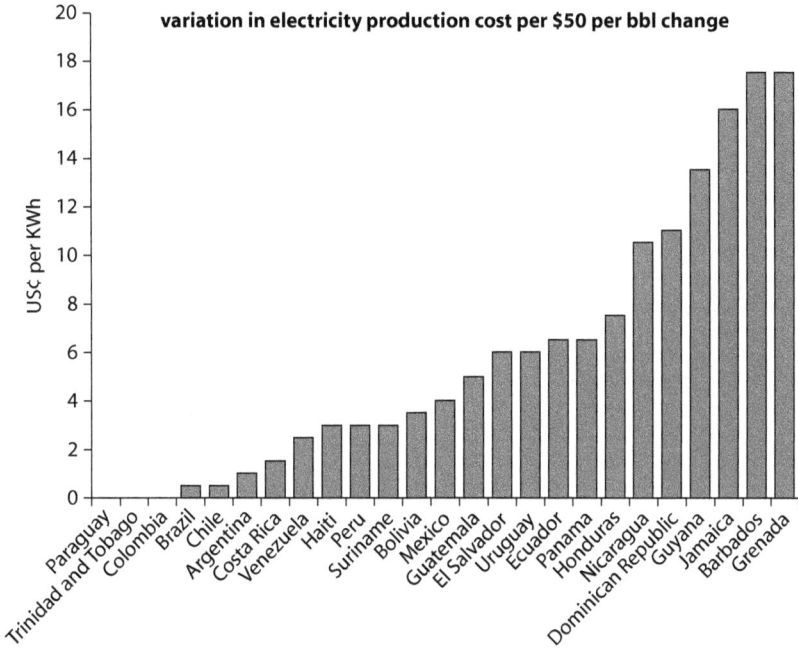

Source: Authors.

Box 4.1

Oil Price Exposure in Guatemala

Guatemala's electricity sector relies on an energy mix consisting mainly of oil, hydroelectric power, and coal. The sector manages volatility by passing prices on to final consumers, with end-user tariffs based on a full pass-through of the whole-sale price. The National Electricity Commission (CNEE), the state power regulator, reviews tariffs every three months, following the General Law of Electricity, which allows cost-increase transfers to final customers. A price stabilization fund, set up under this law, mitigates price hikes for low-income customers. The National Elec-trification Institute (INDE), the state-owned power company, transfers social-tariff funds directly to distributors to subsidize the poorest customers (consumers of less than 300 kWh per month). But a continued upward trend in oil prices will require increasing the subsidy, creating fiscal risk.

(continued next page)

Box 4.1 *(continued)*

The national budget is also affected by the amount of winter rainfall Guatemala receives, which determines how much hydropower can be generated to offset the supply-demand gap when oil prices rise. In 2008, for example, when market prices spiked, the country's winter water supply sufficed to ease the impact of oil price movements. Little rainfall the next winter, however, meant resorting to oil- and coal-fired generation plants, which caused the price of electricity to rise.

The country is working to increase hydropower capacity, which can reduce electricity prices over the medium and longer term. Private-sector generators have used hedging strategies to manage price volatility; however, under the General Law of Electricity, INDE cannot offer the regulatory support required to use financial products as a fund management mechanism.

Source: Authors.

Table 4.1   Electricity Market Structure in Central America and the Caribbean

Vertically integrated monopoly	Wholesale competition	Single buyer with competitive generation
Antigua and Barbuda, The Bahamas, Barbados, Dominica, Grenada, Haiti, St. Kitts and Nevis, St. Lucia, and St. Vincent and the Grenadines	Dominican Republic, Guatemala, Nicaragua, and Panama	Belize, Costa Rica, Guyana, Honduras, and Jamaica

Source: Authors.

wholesale marketers, while the utility maintains the single-buyer role for small consumers. This model is used in several Central American countries. In the **single buyer with competitive generation** model, the utility maintains a monopoly over transmission and distribution, while generation is procured competitively from independent sources, as well as from the utility's own generation. This model is used in five countries of the two subregions.

The more integrated the utility, the less intermediation costs it must absorb. Thus, in countries whose power sectors are still dominated by vertically integrated, state-owned electric utilities, subsidies usually play a more important role. During the 2007–08 rise in fuel prices, governments of many developing countries were led to intervene, either directly or indirectly, in setting fuel or electricity prices (Kojima 2009). The countries best positioned to manage price volatility at a macro level are those in which the government is the majority owner of the transmission system with some level of centralization (i.e., either the vertically integrated monopoly or single buyer with competitive generation model). These countries include Antigua and Barbuda, The Bahamas, Guyana, Haiti, St. Kitts and Nevis, and St. Vincent and the Grenadines.

Power Utility Ownership

Clearly, ownership of the power utility affects the ultimate cost burden of oil price exposure. Utilities could be owned solely by the government or private sector; however, a more common arrangement is some form of shared ownership along the continuum between these two extremes. In most Caribbean countries, the government has the majority share of utility ownership, while the private sector predominates in Central America (table 4.2).

Pricing Mechanisms and Subsidies

The market's electricity pricing policy determines which party bears the ultimate cost burden of oil risk exposure. As previously discussed in chapter 2, a full pass-through of final costs through end-user tariffs means that consumers bear the cost of higher oil prices. Conversely, if end-user tariffs remain fixed as the fuel cost used to generate electricity varies, the utility owner must bear the full cost. In countries with embedded generalized subsidies—ones that extend beyond the poor—the subsidy size relative to fuel-cost increases determines the distribution of higher oil prices between end users and the utility. Since tariffs seldom cover the cost of rising fuel prices fully, many governments transfer funds to their utilities to fill a portion of the gap between tariff revenues and generation costs.

Table 4.2   Majority Ownership of Power Utilities in Countries of Central America and the Caribbean

Government	Private sector
Antigua and Barbuda, The Bahamas, Costa Rica, Dominican Republic, Guyana, Haiti, Honduras, St. Kitts and Nevis, and St. Vincent and the Grenadines	Barbados, Belize, Dominica, El Salvador, Grenada, Guatemala, Jamaica, Nicaragua, Panama, and St. Lucia

Source: Authors.

For most of the countries analyzed in this study, consumers are shielded to varying degrees by tariffs with embedded generalized subsidies. Because tariff increases are often insufficient to cover rising generation costs, the financial position of the utilities may deteriorate. This has been the case for the Dominican Republic, Haiti, and Honduras (box 4.2).

A full pass-through power-pricing policy is in effect in The Bahamas and St. Vincent and the Grenadines—where the government is the majority owner of the power utility—and in Barbados, Dominica, Grenada, Jamaica, and St. Lucia—where the private sector is the majority owner. This policy is sustainable in the long run as it forces consumers to eliminate waste and then seek pathways to improve consumption efficiency. Guyana's move from a partial to a full pass-through regime in 2008 highlights the trade-offs of such decisions on final users and the regulatory implications for managing volatility (box 4.3).

Economic Impacts on the Sector

Government intervention in energy price stabilization has often proven costly or, in some cases, financially unsustainable. In the case of Peru, the government's price stabilization fund has faced challenges to remain

---

Box 4.2

Subsidies and Fiscal Vulnerability in Honduras

In Honduras, the tariff structure and subsidy policies have fully exposed the government to oil price volatility. The pricing mechanisms used to protect consumers from exposure to high and volatile prices, including a stabilization fund and price bands, have affected the budget directly, meaning that the government bears the entire risk of oil price volatility. The national utility, National Electricity Corporation (ENEE), has experienced problems implementing cost-covering tariffs and managing subsidy distribution. In 2008, the subsidy program created a deficit of US$72 million, which grew by another $18 million by 2010. The government's commitment to supporting the electricity sector above a price level of $79 per bbl resulted in $2 billion in outstanding arrears to generators, which the government paid in September 2010. To address ENEE's delicate financial situation, the government raised tariffs, except on residential users who consume less than 150 kWh per month. In 2010, the overall tariff increment was 12 percent.

Source: Authors, with ENEE data.

Box 4.3

Pricing Policy Trade-Offs in Guyana

Guyana initially managed the 2004 rise in oil prices by deferring system maintenance. As the upward march in oil prices continued through 2005, the government approved a 3 percent tariff increase—sufficient to cover the higher cost of
service from the previous year but not enough to cover deferred maintenance.
Not until 2008, when oil prices peaked, was a full tariff review conducted. The
resulting decision for a full pass-through tariff structure meant that tariffs should
have decreased in 2009 as oil prices fell. But they were kept at 2008 levels to overcome the shortfall from previous years and complete system maintenance.
Guyana Power and Light, the state-owned utility, estimated that 4 percent of 2010
tariff revenues, about US$4 million, would be spent on maintenance and new
capacity. Currently, the country lacks the regulatory and institutional capacity to
use hedging instruments. Responsibility for use of financial products rests with
the Ministry of Finance, with oversight by the Public Utilities Commission, the
utility regulatory agency.

Source: Authors.

self-financing amid steadily rising oil prices and has required added fiscal support (box 4.4).

As discussed in chapter 2, the efficient functioning of the power
sector's institutional and regulatory framework depends on the time-
consistent application of directives. Using government price controls and
other nonmarket interventions to accommodate the population's demand
for protection from oil price hikes and volatility runs the risk of weakening the functionality and credibility of the sector's regulatory framework.
Furthermore, once price controls are implemented, it then becomes difficult politically to remove them and return to a market-based regime.

Another management challenge for the power sector is making long-
term generation plans in the face of the uncertainty created by price
volatility. As previously discussed, the planning and building of new
power-generation capacity takes many years to achieve, requiring a
framework for observing the effects of oil price changes on technology
selection. The heightened uncertainty resulting from price volatility can
cause sector planners to delay investments or make inappropriate, sometimes irreversible generation-equipment decisions that affect electricity
costs well into the future.

Box 4.4

Peru's Oil Stabilization Fund

The Fuel Price Stabilization Fund, established in 2004 by Peru's Ministry of Energy and Mines (MEM), aimed at reducing inflationary pressure on the domestic economy by managing the price volatility impact of crude and fuel oil derivatives on domestic consumers. To cushion end users from the pass-through of higher import prices, the MEM put in place a system of price bands for various grades of gasoline, liquefied natural gas (LNG), kerosene, diesel B2, and residual oils (6 and 500) used in the industrial sector. Initially capitalized by the Ministry of Finance with US$57 million, the Fund subsequently received periodic injections, the most recent being $381 million in 2009.

The price-band mechanism set upper and lower limits around which market reference prices (RPPs) could fluctuate. Published weekly by Peru's energy regulatory agency, RPPs correspond to each oil derivative produced by domestic refineries, taking into account international prices, plus freight and other costs. Domestic refineries and fuel importers were to pay into the fund when RPPs fell below the lower band limit; conversely, when RPPs rose above the upper limit, the difference could be used to compensate refineries.

During 2004–06—a period of relatively modest changes in world market conditions—the Fund functioned without major problems. But more recently, particularly in 2008–09, financial management grew more difficult, owing to the need to pay domestic refineries significant amounts and the arrears accrued when the government was unable to make payments.

As of 2009, the Fund had paid refineries more than $1.6 billion, $1 billion of which was paid in 2008 alone. Payment delays led to inefficiencies in the local market, creating long-term problems for producers (in the areas of planning, storage, inventory, and financing). Less availability of operating capital led to quantifiable losses for refineries and, in some cases, consumer supply disruptions. As of April 2010, the government had accrued some $140 million in arrears to local refineries.

Delays in band adjustments also created problems since prices did not adjust to adequately reflect market conditions. In April 2010, the government increased fuel prices, as embodied by the band mechanism, by 7 percent. Prior to that change, LNG prices had been estimated as much as 24 percent lower than market prices and 84-octane gasoline as much as 19 percent lower.

(continued next page)

Box 4.4   *(continued)*

The Ministry of Finance's contingent liability to the Fund became a significant source of instability, affecting the government's annual budget. As a result, by mid-2009, the government began exploring the possibility of hedging exposure to the Fund using a price risk management strategy.

Source: Authors.

Summing Up

Evidence of the management challenges faced by the power sector in heavily oil-dependent countries in the LAC region suggests the types of policy options that could be implemented to reduce vulnerability to high and volatile oil prices. Several structural measures designed to reduce oil consumption are considered in chapters 6–8; before discussing these longer-term strategies, however, we first turn to the various financial instruments that might be applied to better manage price volatility in the shorter term, which are the topic of the next chapter.

Note

1. Oil comprises only 5 percent of the world's power generation, compared to 13 percent for the LAC region.

Reference

Kojima, M. 2009. *Government Responses to Oil Price Volatility: Experience of 49 Developing Countries*. Extractive Industries for Development Series #10. Washington, DC: World Bank.

CHAPTER 5

Price Risk Management Instruments

Oil price volatility can be managed over the shorter term (e.g., 1–2 years) using price risk management instruments. Commonly referred to as hedging, such tools can reduce the uncertainty associated with commodity-price volatility, particularly its impact on national budgets. The aim is to manage existing price exposure, which is generally a function of current structural conditions.[1] However, hedging instruments should not be used as a substitute for more basic structural measures designed to reduce oil consumption (chapters 6–8).

This chapter begins by summarizing the variety of instruments used to manage price volatility, along with an overview of their relative costs and benefits. It follows with a review of the risk assessment process, an important prerequisite for appropriate use of the instruments. Since price exposure is created by the relationships and transactions between various power-sector actors, the chapter notes the importance of considering the interaction between private- and public-sector actors and mechanisms. Finally, recommendations are offered for strengthening institutional frameworks to support commodity risk management programs. All of these issues are evaluated from the perspective of governments of energy-importing countries concerned with managing the short-term budgetary uncertainty created by oil price volatility.

Overview of Instruments

There are two main categories of price risk management instruments: (i) physical and (ii) financial. Physical instruments can include storage, strategic timing of purchases and sales (e.g., "back-to-back" trading), forward contracts, minimum/maximum price forward contracts, price-to-be fixed contracts, and long-term contracts with fixed or floating prices. Financial instruments include exchange-traded futures and options contracts and over-the-counter (OTC) products, including swaps, collar contracts, customized options, commodity-linked bonds or loans, trade finance arrangements, or other commodity derivatives. Deciding which instruments to use to manage volatility depends heavily on the specific risks an entity is trying to address and the structure of commercial relationships and thus the financial issues that impact that entity.

Physical Instruments

Physical price risk management involves contractual negotiations between buyers and sellers regarding the terms under which exchange of the physical good will occur. Managing price risk through physical instruments can include the following:

- *Storage.* Many countries use strategic reserves to protect themselves against the risk of price/supply shock. Solid infrastructure, well-managed facilities, transparency about volumes and costs, and a rules-based system for managing draw-downs and replenishments are important components of an efficient reserve system.

- *Strategic timing of purchases and sales.* This simple, conservative way to manage price volatility works if there is sufficient flexibility in the ability to set contractual terms. One common mechanism is "back-to-back" trading, which refers to the ability to set the terms and timing of the purchase with those of the sale of the good, which uses the commodity as an input. Price risk is minimized because there is little time lag or difference in terms between the purchase and sales agreement. For example, a private generator buying inputs priced on the basis of a specific date (or averaging period) of a price index can efficiently hedge the price risk by selling power using a pricing formula based on the same underlying specific date (or averaging period) of that index.

- *Forward contracts.* Forward contracts are agreements to purchase or sell a specified product on a specified forward date for a specific, predetermined price. Forward contracts require physical delivery of the product, and payment is expected to occur at the forward delivery date. The seller of a forward contract has no knowledge of the prevailing market price at the time of delivery but agrees to a specified, predetermined price ahead of the delivery date, thus absorbing the price risk on behalf of the buyer.

- *Minimum/maximum price forward contracts.* Though rarely used in energy trading, min/max forward contracts are used in other commodity sectors. They provide a minimum or maximum price cap or ceiling, which is negotiated at the time of the contract. From a buyer's perspective, one advantage of such contracts is the ability to take advantage of the lower price if the prevailing market price at the time of delivery is lower than the predetermined maximum price. If the prevailing market price at the time of delivery is higher than the predetermined maximum price, the buyer has a guaranteed maximum purchase price, and is not obligated to buy at the higher market level.

- *Price-to-be fixed contracts.* These contracts allow the buyer to negotiate flexibility in the contract, which allows fixing of the underlying price basis at a time decided by the buyer.

- *Long-term contracts with fixed or floating prices.* These are variations of the above, in contracts with longer maturities.

One major advantage of embedding risk management solutions in physical supply contracts is simplicity; that is, physical contracts are customized to provide the required price protection, and there is no need to manage an additional counterparty relationship (e.g., with an investment bank providing a stand-alone financial hedge) or additional documentation. Physical hedges can generally be negotiated with existing suppliers and are somewhat easier to manage from an accounting and auditing perspective. For governments, on the other hand, physical hedging can be complicated since importers may be private actors not directly involved, from a supply-chain perspective, in the policy mechanism (e.g., stabilization or subsidy program) that is creating financial risk for the government.

Financial Instruments

Financial risk management products are purely financial contracts negotiated separately from the physical supply of the actual commodity. They are available as (i) exchange-traded products (i.e., through established commodity futures exchanges) or (ii) OTC products (i.e., contracts traded between two independent counterparties). Investment banks and multinational trading companies can act as the counterparty to these types of transactions.

Exchange-traded products. These products are traded on commodity exchanges, which act as clearinghouses that transfer risk from one commercial participant to the other. Commodity exchanges perform functions in price formation and provide transparency to the market. They also perform a credit/counterparty risk management function for the market since all trades going through the exchange are backed financially by the exchange itself. Commodity exchanges offer the following types of contracts:

- *Futures contracts.* Futures contracts are similar to forward contracts in that they are agreements to buy or sell a specific quantity of a commodity at a specific price on a specific future date. Unlike forward contracts, however, futures contracts do not necessarily require physical delivery to fulfill the contract. Futures contracts can be considered "paper" contracts because they can be settled without physical delivery; they provide the advantage of being able to "lock in" a purchase or sale price in advance of the product delivery. This is beneficial when prices are at a level that covers costs or is a financial break-even point.

  Futures contracts are generally used in parallel with activities in the physical market. For example, a buyer could use a futures purchase to lock in a price in advance of the physical delivery; when the time arrives for physical delivery of the goods, the buyer would use a futures sale to essentially "sell back" the obligation to buy on the exchange. The gain or loss on the futures (financial) transaction would then be offset by a roughly equivalent gain or loss on the physical transaction. A major disadvantage of hedging with futures, however, is that these contracts create unknown, unpredictable contingent liabilities for the hedger. Using the above example, in cases where the market had moved down in between the two transactions, a buyer using futures would have to pay the market counterparty the difference between the futures purchase and sale, which could be a sizable sum. The inherent credit risk in trade of these contracts means that a hedger using futures must be prepared not only to make these payments when the contracts settle, but also to post collateral to the market counterparty throughout the life of the contract.

- *Options contracts.* Options contracts provide the opportunity, but not the obligation, to buy or sell a specific quantity of a commodity at a specific price on a specific future date and can therefore be used to cap prices by creating a price ceiling. They are purchased by the hedger in much the same way as insurance is purchased; the buyer purchases the right, but not the obligation, to declare a futures contract. The instrument is valued for the added flexibility it offers. If market prices are trending in a positive direction, there is no need to exercise the option, meaning that the hedger can avoid locking in a fixed price level, as is done with a futures contract.

  The two types of options contracts are puts and calls, which are agreements to either sell (put) or buy (call) a futures contract at an agreed-on strike price and future expiration date. Both contract types have a cost or premium, based on the relationship between the strike price and the current market price, the time between purchase of the instrument and its expiration date, and market volatility.

OTC products. Over the years, the need to customize financial risk management tools to meet specific needs of market participants has resulted in greater use of customized and OTC products. In response to increasing standardization of products, the need to better manage credit exposures, and regulatory reform, some of the products have been moving on to exchange platforms. Customized and OTC products include the following:

- *Swap contracts.* These financial transactions are designed to manage exposure to two financial streams over a period of time. In a simple swap contract, one leg of the contract is fixed while the other is floating. Contracts can be structured to have an automatic settlement of the difference between, for example, the fixed and floating prices.

- *Collar contracts.* These contracts combine put and call options to limit price exposure to fluctuations within a specific band or "collar." The buyer has both a price ceiling and floor, achieved by simultaneously buying a call and selling a put at two price levels that create a price band, within which the buyer takes the market price. If the price moves above the price ceiling, the buyer does not pay more than the agreed ceiling price. If the price moves below the price floor, however, the buyer has a liability and must pay the difference between the price floor and the lower market price. The overall premium for a collar can be significantly lower than for a stand-alone call option since the buyer is simultaneously buying a call and selling a put. Such contracts are sometimes marketed as "costless;" however, it should be cautioned that prospective hedgers must be aware of and have the ability to manage the liabilities they could potentially incur if prices were to move below the floor level.

- *Customized options.* An example is an Asian option, which settles automatically over an average time period rather than at a specific, monthly expiration date.

- *Commodity-linked bonds or loans.* These more complex types of financial transactions are often constructed to help mitigate the exposure of investment projects or manage debt related to commodity activities. As with other commodity instruments, the cost of a commodity indexed loan depends on the tenor or time frame of the price protection. For example, the cost of 10-year price protection is higher than for 3-year protection. Cost also depends on the relationship between the price level protected and the current market price, as well as the degree of market volatility.

OTC contracts are governed by internationally recognized agreements, called International Swaps and Derivatives Association Master Derivatives Agreements or ISDAs. There is counterparty risk on these contracts since either party could default. Experience shows that governments using financial instruments for purposes other than managing commodity exposure may, in some cases, encounter legal or administrative barriers to their use. These may include the following:

- Explicit restrictions limiting the government's ability to (i) enter into ISDA-based Master Derivatives Agreements, (ii) finance risk management strategies (e.g., by using the national budget to pay premiums), and (iii) transfer funds to overseas market counterparties in order to support risk management transactions.

- Explicit restrictions limiting the ability of a state-owned enterprise to enter into hedging transactions.
- Legal frameworks that, by omission, do not allow the government to use derivatives contracts because they are not included in the list of approved financial instruments.
- Budget frameworks that, by omission, do not allow the government to finance derivatives contracts.
- Audit controls that do not create appropriate support for the use of and expenditures associated with financial risk management instruments.
- Administrative constraints concerning disagreements or lack of clarity about the respective roles of (i) various ministries in establishing energy sector–related policies and (ii) public- and private-sector actors with key sector functions (e.g., importing).
- Administrative constraints related to the inability to account for risk-management expenditures (payments and receipts).

Addressing these barriers is key to developing appropriate institutional frameworks to support commodity risk management, an issue discussed later in this chapter.

It should be noted that no single instrument is superior at managing volatility. Deciding which instruments to use should be based on a careful assessment of the specific financial risks and an evaluation of the benefits versus the costs/risks and constraints associated with using specific tools (table 5.1).

Finally, users of price risk management instruments should be aware of the several types of risks not covered by these products that they themselves could generate. First, since commodity risk instruments are generally traded in U.S. dollars, their use may create the risk of fluctuation between the U.S. dollar and the local currency. Second, credit risk may arise since settlement of these contracts results in financial obligations to the counterparty. Third, basis risk can arise when the price index of the hedge contract differs from that of the actual physical commodity and moves in divergent ways. For example, though the prices of many oil derivative products (e.g., heavy fuel oil [HFO] and diesel fuel) are highly correlated with crude oil, and while the West Texas Intermediate (WTI) and/or Brent crude oil contracts are generally considered the reference price for many energy derivatives, price movements in the respective markets can differ; as a result, the degree of correlation between price movements can change over time.

Table 5.1 Overview of Selected Hedging Instruments: Advantages and Disadvantages

Product	Interest of power-sector agent	Benefits	Costs/risks/constraints
Forwards	Integrating price risk management solutions into physical supply contracts.	Since forwards are physical supply contracts, the risk management solution is embedded in the supply contract, and there is no need for a separate contract/ documentation. Pricing of forward contracts can be customized to the needs of the hedger: prices can be fixed, floating, or include caps/floors and collars (a pre-agreed range or band). Depending on the pricing formula used, forwards have the same benefits as the financial products described below.	May be complex for government to implement if it is not directly involved in physical importing. Depending on the pricing formulas used, will have the same costs/risks/constraints as the financial products described below.
Futures	Establishing fixed price certainty without interest in taking advantage of future upside or downside price movements.	There are no upfront costs. It is possible to lock in forward prices through a financial contract.	Locks in fixed prices and limits the hedger's ability to take advantage of positive price movements that may occur in the future. Creates unknown and unpredictable future liability since hedger will owe the market counterparty if the market moves in an adverse direction. Requires financing of a credit line or a credit guarantee. Requires managing cash flow/liquidity requirements to support (potential) daily margin calls.

Instrument	Purpose	Description	Considerations
Options	Establishing a cap or floor on prices but maintaining flexibility to take advantage of lower or higher prices that may occur in the future.	The hedger can lock in maximum and minimum prices and take advantage of positive price movements that may occur in the future.	Has an upfront cost, or premium, pricing for which is market-driven and can be volatile. On an indicative basis, premium costs but can range from 5 to 12 percent of the value of the underlying price for 6–18 month coverage.
Collars	Establishing a price band or range.	Price exposure is limited to a price band (collar) that has both a ceiling and a floor. The upfront costs can be lower since the hedger is, for example, simultaneously buying a call option and selling a put option.	Creates unknown and unpredictable future liability since hedger will owe the counterparty if the market moves below the price floor. Requires financing of a credit line or a credit guarantee. Requires managing cash flow/liquidity requirements to support (potential) daily margin calls.
Swaps	Establishing price certainty without interest in taking advantage of future upside or downside price movements.	There are no upfront costs. As with futures contracts, swaps can be used to lock in fixed price levels. Swaps provide the ability to simultaneously manage two commodity exposures or financial flows.	Creates unknown and unpredictable future liability. Requires financing of a credit line or credit guarantee. Requires managing cash-flow requirements to support (potential) daily margin calls.
Commodity-linked bonds or loans	Combining price protection into a loan so that repayment obligations are lower when prices move in an adverse direction.	On a more macro level, these instruments could be used to connect borrowing or financing programs to the performance of a specific commodity index.	Can be more complex to structure. May not be effective as a hedge for specific, short-term commercial exposures.

Source: Authors.

To illustrate, during the past decade, the respective prices of HFO and WTI diverged markedly in 2004–08, subsequently rising in step in 2009–10 (figure 5.1). The correlation between HFO and crude oil during this time frame was 0.960 in levels and 0.715 in logarithmic differences (approximately equal to percentage changes), suggesting that the WTI contract may have served as a reasonable hedge for HFO.

Risk Assessment

Careful risk assessment is a critical step for any entity considering using price risk management instruments. While many countries are net importers of fuel and thus exposed to the risk of upside price movements, further analysis is needed to more specifically isolate and quantify the nature of that exposure.

A more detailed assessment of commodity price exposure would require two main components: (i) a supply-chain risk assessment that defines the roles and responsibilities of each actor in the sector, describing how each is affected by price volatility and (ii) a financial risk assessment that quantifies the price exposure resulting from specific commercial transactions or policy interventions and decisions.

Figure 5.1   Prices of WTI Crude Oil, Heating Oil, and HFO, 2001–10

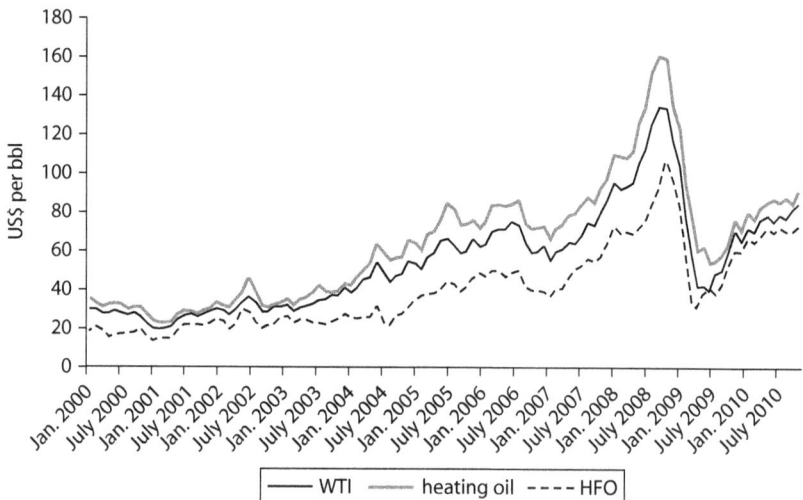

Source: Gilbert 2010.

Supply-Chain Risk Management: Example for the Power Sector

An example of a typical supply-chain risk assessment for the four main actors in the power supply chain—generators, distributors, consumers, and government—can be described as follows:

- *Generators*, which may be privately owned and are often the main importing agent, have a straightforward and direct exposure to commodity prices. If the prices of commodity imports increase, costs will increase and need to be passed on to distributors in the form of higher power prices so that generators prevent financial losses. In some countries, sales contracts that govern the commercial relationship between the generators and distributors pass on market prices from the former to the latter. Generators may run a short-term (e.g., one month) risk since the commodity component of these price formulas may lag; but they can manage this exposure independently, perhaps by negotiating import contracts that mirror the lag.

- *Distributors*, whether government controlled or not, often purchase power from the generators and sell to consumers under a complex system subject to many risks, only one of which is exposure to price volatility. Ideally, distributors should buy power from generators on the basis of formulas that mirror those used to sell power to consumers. This would allow for back-to-back trading, which creates little price exposure for the distributor, who is merely acting as a market intermediary. In theory, the pricing structure in some countries attempts to provide this financial protection to distributors since the tariffs are designed to allow for pass-through of market prices. In practice, however, distributors have at times struggled not only with appropriate implementation of tariff changes, but also with the financial loss associated with other serious problems. The lag of the price pass-through in the tariff system can result in financial losses for the distributors, in turn, contributing to overall inefficiencies and underinvestment.

- *Consumers*, who purchase power from distributors, are often cushioned from price volatility since tariffs are not always re-aligning to market price levels. When market prices increase, tariff adjustments often occur only after the government has suffered substantial losses and can no longer face the financial burden of subsidizing prices. When market prices decrease, tariffs have not always decreased equivalently.

- *Government* often has financial exposure to the power sector in the form of direct support to distributors. In some countries, this can represent a significant portion of the national budget and is a contingent liability that has been difficult to predict. The government is often concerned with two major exposures: (i) the impact of rising commodity prices on the cost of production and pass-through of these costs to consumers and (ii) the risk of financial losses associated with un-hedged price exposure within the system, which is more serious for distributors unable to pass through price increases to consumers. In some cases, the government's contingent liability to the sector is used as compensation for financial losses of the distributors.

Financial Risk Management: Example for the Power Sector

Once the roles and responsibilities of the actors in the power-supply chain and the ways they are affected by price volatility are understood, financial risk assessment can help to quantify the specific price exposures faced by actors in the system. Financial risk assessment of commodity price exposure is based on careful identification of product, price level, volume, and duration, as follows:

- *Product.* Across the region, the mix of energy products used to produce electricity varies by country and company and can change over time. Typically, energy production in the region is based on consumption of HFO, diesel fuel, natural gas, and coal. Since the market movements of each of these products can differ, it is important to specify the appropriate product and product mix and monitor how they change over time.

- *Price level.* This is the price basis at which inventories are valued and purchased and sales commitments are made at every step in the supply chain. Pricing formulas used by various entities across the region can be complex and must be analyzed carefully in order to understand the degree to which the price calculated by these formulas is sensitive to the underlying commodity index. Depending on which risk is to be hedged, this sensitivity analysis is particularly important for evaluating the commodity price impact on (i) purchases of energy commodities by the importer and/or generator, (ii) sales contracts that determine the price of power sold by generators to the distributors and/or tariffs charged to consumers, and (iii) the financial impact of the interactions between those contractual agreements.

- *Volume.* Price exposure is also affected by the volume of purchases of various fuel products and the volume of electricity sales to consumers, both of which can change over time. For example, a generator might sell to distributors using a formula based on the HFO product, when, in fact, the energy was produced using a different type of fuel (e.g., natural gas or hydropower). The volumes and mix of commodities used in generation and dispatch, therefore, can differ from the volumes and mix of commodities used in the pricing formulas that govern sales to distributors.

- *Duration.* At the commercial level, the time period for which an entity is exposed to the risk of an unfavorable price movement is important. For example, the duration of the price exposure can be determined by how the commodity index component of the pricing formulas passes through from the generator to the distributor and from the distributor to the consumer. Depending on the timing of the pass-through, the duration of exposure can range from just 1–2 weeks up to 8–12 months or longer if prices are not adjusting with the market. From the government's perspective, the duration of price exposure may depend on the time frame for which a particular policy (e.g., a subsidy program) has been committed. In many cases, the government's most serious exposure concern is short term (e.g., the annual budget cycle and related commitments to specific programs or investments).

Hedging Possibilities for Power Market Agents

Since electricity sectors in countries across the LAC region are frequently managed by a complex set of interactions between public- and private-sector entities responsible for various aspects of the supply chain, it may not be appropriate to design a hedging strategy based on a simple calculation of the volume of imports of the primary commodity. A more careful financial analysis of the product mix, volume, price level, and duration of the exposure, particularly as it relates to the government's exposure, is important.

Also, given the complex nature of commercial relationships in the power sector, risk assessment is important since non-price risks to the system may need to be isolated and managed independent of a price risk management strategy. These may be (i) regulatory risks, which are associated with managing the contractual relationships between generators, distributors, and consumers and the applicability of electricity tariffs with pass-through mechanisms; (ii) contractual or operational risks, related to the inability to recover costs from consumers; or (iii) credit risks, related to a buyer's inability to obtain credit terms that would help to optimize purchase strategies.

Although risk assessment is needed to determine hedging strategies that best fit specific cases, hedging possibilities for specific market agents can be described in general terms. Generators, for example, can have short-term price risk created by the time lag between the point at which fuel purchase prices are set and the point at which electricity is sold to distributors. If they first determine the terms of a sale of electricity to distributors, they will have a short position, meaning they are exposed to the risk of a rise in generation costs before the terms of fuel purchases can be finalized (box 5.1).

Generally, this scenario has two hedging possibilities. The first is to buy oil futures at the point when the terms of an electricity sale to distributors are confirmed, and then to unwind the futures position when the terms of a physical fuel purchase are finalized. The second is to lock in long-term fixed forward contracts when oil prices allow for a reasonable return.

Distributors are exposed to price risk if they are unable to pass on purchase costs to consumers. This creates an inefficient price exposure that is difficult to hedge, particularly if agreements for purchasing power

---

Box 5.1

Hedging Natural Gas for Mexican Generators and Final Consumers

In 1995, PEMEX Gas, Mexico's state-owned enterprise responsible for commercializing the sale of natural gas in the country, and the Federal Electricity Commission (CFE), the state-owned power generator, agreed to start using financial instruments that year to reduce CFE's price risk. With the opening of the natural gas market two years later, PEMEX Gas began offering risk management services to distributors and industrial clients (mainly large consumers, such as steel, glass, and cement producers). By 1998, 80 percent of CFE's gas consumption was hedged, along with 22 percent of the distributors' volumes and 8 percent of industrial users' volumes. Over the past decade, the value of hedged positions grew dramatically—from US$1.45 million in 2000 to $228.13 million in 2010.

Source: Authors, with PEMEX and PEMEX Gas data.

from generators and selling to consumers are fixed or regulated in a way that disallows price movements that follow market trends (box 5.2).

Because the problem is structural, solutions for this scenario are limited to either (i) re-assessing pricing mechanisms that govern the purchase of power from generators to align more closely with sales pricing

---

Box 5.2

Hedging Natural Gas for Mexican Distributors

In October 2003, Mexico's Energy Regulatory Commission issued a directive allowing natural gas distributors to use financial instruments to hedge against sharp price movements. Following a period of heightened volatility, and at the distributors' request, regulators agreed to allow the resulting hedged fixed prices to be reflected in the final price for small consumers (amounting to 360 giga-calories per year, in addition to the international reference). This mechanism permitted the final price structure to reflect price components for the fuel and its transport, storage, and distribution; along with an adjustment to reflect the hedged price, if applicable.

Under the terms of the initial directive, distributors had discretion over hedging counterparties as long as the hedge provider could demonstrate two years of experience, legal registration, and financial viability to meet obligations arising from the hedging contract. Also, distributors themselves could offer hedging products to larger consumers. A final agreed-to change in the pricing formula allowed distributors to pass on the financial costs of the hedging transactions, together with the agreed price in the hedging instrument. As a result, consumers got a pass-through of the hedged price, which reduced the volatility of their bill. In this case, the distribution company was completely neutral to the hedging strategy as the price risk was transferred to consumers through the hedged price, with the financial cost of the instrument fully paid by consumers.

In July 2007, this directive was modified to support a pooling mechanism through PEMEX Gas that allowed for a homogenous hedging strategy for all consumers in the pool. To achieve better contractual and price conditions, the volumes of distributors across geographic regions were aggregated together. Some distributors were hesitant to get a specific hedging instrument lest they might become uncompetitive if another distributor got a better hedged price. Once again, the hedged price and the financial costs of the hedging transactions were passed on to final consumers.

Source: Authors, with PEMEX and PEMEX Gas data.

mechanisms or (ii) re-assessing pricing mechanisms that govern sales to consumers to align more closely with purchase pricing mechanisms.

Consumers in many countries are cushioned from price volatility since tariffs are not always re-aligning to market price levels. Large-scale consumers could hedge by purchasing power from agents on a capped price basis, which would require paying a premium to ensure that prices do not rise above a pre-agreed level. Small-scale consumers could be supported by developing a hedged subsidy program that uses call options contracts to create a price cap. If done on a financial basis, payouts from the market instrument could be used to fund the subsidy program when prices move above the price cap.

Government exposure manifests in the form of contingent liabilities to agents that suffer financial losses as a result of price volatility and/or to subsidy and stabilization programs. Generally, governments are concerned about severe price shocks, which create unmanageable increases in these contingent liabilities (box 5.3).

Box 5.3

Evolution of Panama's Oil Hedging Strategy

Each year, Panama consumes millions of barrels of HFO, used for domestic power production. In 2009, fiscal transfers to the government-managed Tariff Stabilization Fund (TSF), a subsidy program to mitigate the impact of oil price volatility on consumer costs, totaled US$96 million; by mid-2010, $66 million had been budgeted for that year. The government has a growing concern over the TSF's contingent liability since the fund fluctuates with the market price of fuel oil. Consumer power prices are based on established formulas, and are forward adjusted by the energy regulator and distribution companies every six months. A fixed fuel-oil price, estimated annually, is used to allocate funds to support the TSF. If market prices move above the fixed price reference, the government faces a financing problem since the subsidy program costs will exceed those originally budgeted.

Recognizing the importance of establishing a long-term policy for this type of transaction, the government initiated the National Strategy for Hedging the Risk of Hydrocarbons, which was approved by cabinet resolution 157-A on December 15, 2009. Designed to ensure that the government can fix its budgetary support to the TSF with more certainty, the strategy uses a price ceiling for price insurance.

(continued next page)

Box 5.3 *(continued)*

If prices move above the fixed price ceiling, the contract pays the difference between the strike and market prices. In November 2008, the government purchased an option for a 12-month period to support the TSF for 2009. The option strike price, at the time, was estimated at 100 percent of the country's consumption. The option was purchased out of the money when the market price was quite low. During the contract period, market prices moved higher than the strike price and the government received a payout of $19.4 million, net of the premium cost. But this amount was not enough to cover the TSF's total needs since low rainfall that year impeded hydropower generation, increasing reliance on thermal generation above the total consumption originally estimated.

For 2010, the government estimated higher fuel-oil consumption and decided to hedge this exposure in two phases using two distinct options contracts. In the first contract, executed January 4, 2010, the government hedged 65 percent of estimated annual consumption (about 2.7 million barrels). The strike price for the first options contract was equivalent to the fuel oil price used by the energy regulator and the distribution companies for the first six months of 2010. At the time of the transaction, this strike price was already in the money (i.e., equivalent to prevailing market prices) and the premium cost for the option was $10.17 per bbl, resulting in a total cost of $26.9 million. This first contract covered the January–December 2010 period. As oil prices remained below the strike price over most of this period, the government received $0.9 million in total gross payouts from this contract.

On May 26, 2010, the government took advantage of a market window to buy a second option for 2.5 million barrels, covering 35 percent of the estimated consumption for 2010 (thereby closing the second phase of the 2010 hedging strategy) and 25 percent of the estimated volume for 2011 (the first phase of the 2011 strategy). The strike price for this contract was $71 per bbl, matching the fuel-oil price established by the energy regulator and distribution companies for the second half of 2010. On the day of the transaction, the strike price was out of the money by about 14 percent (i.e., above prevailing market prices).

Ministry officials have explained the hedging mechanism to other government agencies and established that its objectives are not speculative. From the government's perspective, although the premium cost for the first phase was somewhat high, it was a worthwhile investment in light of the benefits gained by budget certainty. Currently, the government is in the process of expanding the hedging program to cover liquefied petroleum gas (LPG) for domestic household consumption and diesel fuel for public transport.

Source: Authors.

Simulations can assist in evaluating alternative hedging strategies and can be done on a backward- or forward-looking basis. The example presented in box 5.4 is not intended as guidance for specific hedging strategies; rather, it demonstrates the type of technical analysis that can be performed using information about specific price exposure of an actor in the supply chain.

Institutional Frameworks for Commodity Risk Management

In the 1990s, the financial crises in East Asia and Latin America drew attention to the quality of public debt management in developing countries and its role in reducing developing countries' vulnerability to crisis. Similarly, the 2008–09 food and fuel crisis and subsequent increased market volatility have drawn attention to the extent to which developing countries are exposed to commodity price volatility. Discussions with governments about these problems have revealed a gap in technical capacity and knowledge. Governments considering the use of commodity hedging tools require similarly sound frameworks for their use as for other issues of fiscal policy.

Many countries have only a partial understanding of the specific exposure to commodity price risk, yet the details about how and where it affects the national budget are critical, as is coordination of

---

Box 5.4

Simulating Alternative Hedging Strategies: A Simple Example

A simple simulation might be conducted on three sample hedging strategies evaluated against market movements between 2003 and 2010. The strategies could include a futures-based strategy using WTI, diesel fuel or natural gas futures, or an options-based strategy using WTI.

After selecting the instruments, the results can be evaluated using such criteria as (i) the average cost paid for electricity, (ii) the standard deviation of this cost, and (iii) a measure of the "spikiness" of these costs (i.e., the one-sided standard deviation of the positive changes in these purchase prices over the duration of the hedge).

(continued next page)

Box 5.4 *(continued)*

In this case, a simple WTI options hedging strategy is more effective than an unhedged or a futures hedge strategy at reducing the average price paid for electricity, the price variability, and susceptibility to price spikes (box figure 5.4.1).

While hedging instruments cannot reduce exposure to price trends, they can be used to cushion the impact of extreme price movements, such as the price shock illustrated here.

Box Figure 5.4.1   Simulation of WTI Hedging Scenarios

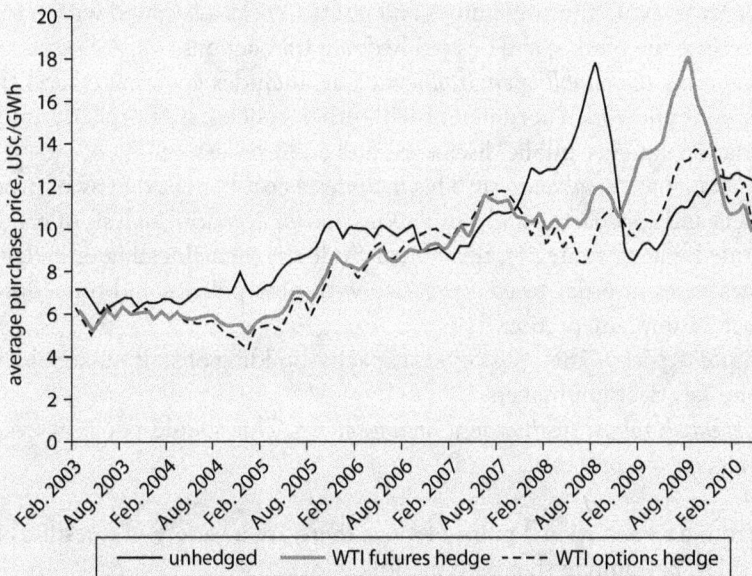

Source: Authors.

the risk management strategy with other policies and investment strategies over time. Once these issues have been discussed, the technical design of a hedging strategy can be implemented, although the process may be challenging in environments that have limited experience with derivatives.

Identifying and managing the trade-offs between expected cost and risk are an important component of a hedging strategy design, as is assigning roles and responsibilities for the institutions involved. The key steps in the overall process of establishing a commodity hedging strategy are as follows:

- *Assess risk*. This step identifies the risk composition (products, price level, time frame, and volume); it also defines a base case and evaluates it against a variety of market scenarios.[2]
- *Document objectives*. This is done by establishing the realistic limits of the approach, obtaining consensus from a broad range of stakeholders, and clearly communicating what the hedge strategy is and is not designed to do. This may include clearly establishing the price to be defended (i.e., through a link to price levels assumed in the budget) in order to avoid the ambiguities and political risks associated with trying to time the market and desired hedging transactions.
- *Evaluate the enabling environment*. This includes governance and the legal framework, coordination with other policies, staff capacity, information systems, public disclosure, and audit processes.
- *Conduct technical analysis*. This includes a cost-benefit review of products and approaches and simulations and/or scenario analysis of prototype hedging strategies; this may include the virtual testing of hedging strategies in order to gain practice with new policies and procedures before implementation.
- *Build capacity*. This step covers capacity-building of staff, stakeholders, and key decision-makers.
- *Establish robust institutional arrangements*. This should occur at every stage in the process.

From an institutional point of view, there are a variety of questions to consider; key among them are the following:

- Which entity/government officials have been empowered by law to make decisions regarding the hedging strategy, enter into the hedging transaction, or manage payments and receipts from a hedging transaction?
- Which units should develop the proposed hedging strategy and which ones should approve it? Which ones are responsible for program audit and control?
- How will decisions be documented (e.g., in meeting minutes or a formal strategy document)?

These issues are critical to obtaining program consensus, for program governance, communication, and marketing, and for protecting program integrity and defending it if it comes under criticism, either politically or financially, at a later stage (box 5.5).

The institutional arrangements necessary to support commodity risk management may vary by country but should always involve formal documentation of the policies and procedures that support decision making, resource allocation, implementation, and monitoring of a commodity risk management strategy.

Box 5.5

Mexico's Oil Hedging Strategy: Institutional Capacity for Risk Management

Mexico's state-owned oil company, PEMEX, the world's third largest oil production company, pays taxes and levies totaling about 60 percent of sales. This represents more than one-third of overall government revenues, meaning that the federal budget is vulnerable to oil price declines. Unhedged price volatility impedes expenditure planning and thus the financing of social expenditure programs.

In response, Mexico's Ministry of Finance and Public Credit has implemented an oil price hedging program—part of a three-pronged, public finance strategy to guarantee sustainability, including adequate liquidity and financial risk management.

Each year, Mexico's Congress establishes a projected oil price for budgetary calculations, based on a pre-established formula using historical and futures prices. The Ministry of Finance and Public Credit designs and executes the oil hedging strategy based on the projected price, with funding from the Oil Revenues Stabilization Fund (FEIP), created in 2001. The hedging strategy is agreed to, based on discussions between the finance minister, under-secretary of finance, and deputy under-secretary for public credit. Any revenues obtained from the hedging transactions are used to offset lost revenues from price declines that have adversely affected oil marketing and sales.

Currently, the government implements the hedging program using the purchase of put options, giving it the right, but not the obligation, to sell oil in the future at a pre-determined strike price equivalent to the projected price set in the budget for the next year. This strategy creates a floor in the price for oil exports, giving the country an opportunity to take advantage of upward price movements should they occur. Typically, the government transacts 12-month, put-option contracts, with a strike price equivalent to the projected oil price used to develop the national budget.

(continued next page)

Box 5.5 *(continued)*

Institutionally, the Ministry of Finance and Public Credit develops the hedging strategy and purchases put options for the following year; it pays the premium from the FEIP via its financial agent, the Central Bank, which runs a competitive process for each transaction to determine market counterparties. In the early program years, Mexico hedged 20–30 percent of its net oil exports. In 2008, following a sharp price increase, the total amount of net oil exports was hedged. For the 2009 fiscal year, 330 million barrels, or 100 percent of net oil exports were hedged via the purchase of put options, with a strike price of US$70 per barrel. The total cost of the hedge was $1.5 billion. At the end of 2009, the options contracts settled with a payout of $5.085 billion. For the 2010 fiscal year, about 222 million barrels or 60 percent of oil exports were hedged through the purchase of put options, with a strike of $57 per barrel, below the official projected price of $65.40 per barrel, on which Mexico's 2011 budget relies. The premium cost was $3.66 per barrel, and the total cost of the hedge $1.172 billion. For 2012, the government hedged 211 million barrels using put options to protect a price level of $85 per barrel.

Mexico's oil hedging program is now nearly a decade old. A strong set of institutional arrangements has provided a solid foundation for the government's risk management strategy. Its systematic approach has resulted in increased sophistication and capacity. The government has a clear and consistent message, which it takes care to communicate to the public: The objective of the hedging strategy is not to profit directly from a fall in the price of oil, but to hedge the existing financial risk that Mexico faces, owing to its heavy dependence on oil revenues.

Source: Authors, with PEMEX data.

Some important fundamentals to keep in mind while establishing institutional arrangements include the following:

- Governance structures need to be in place to support both the decision stage (objectives, budgets, and choice of instruments) and execution stage (transactions, reporting, and controls) of the process.
- Operational policies and procedures need to be carefully developed to ensure control, transparency, and clear authorities for managing relationships with market counterparties; execution and recording of transactions; and payments, receipts, and settlements.

- Since markets change on a daily basis, it is important to establish procedures for ongoing monitoring and reporting. Derivative transactions are generally monitored daily, using a mark-to-market process that provides information on the value of the hedge against current market conditions.
- Depending on the complexity of the strategy, the use of certain tools may require establishing a technical infrastructure (trading desk and information technology systems to manage transaction reporting and monitoring).

At a high level, it is often helpful for governments to establish a steering committee to govern all aspects of the commodity risk management strategy. The steering committee may comprise a broad array of stakeholders, ranging from representatives of finance, treasury, and energy ministries, other public-entity representatives with jurisdiction over components of the strategy (e.g., central bank and/or market regulators), and private-entity representatives directly or indirectly affected by the risk management strategy. The steering committee should take an active role in formalizing institutional policies and procedures during the strategy's development phase and, once in place, provide ongoing oversight and monitoring.

The steering committee should also establish a budget for the hedging strategy, taking into consideration the fixed costs and potential contingent liabilities (if any) associated with use of a specific instrument. For example, the use of a call option, which would effectively cap the price of imports, would require fixed costs in the form of a premium, which may need to be paid upfront at the time the transaction is booked. Since identifying a source of funding can be challenging, countries with limited liquidity may wish to consider borrowing to fund the hedging program. The Caribbean Catastrophe Risk Insurance Facility, which provides insurance against the risk of hurricanes and earthquakes, provides a precedent, given that some countries used World Bank funding to finance participation. Similarly, in the agriculture sector, a World Bank's International Development Association (IDA) project in Africa was restructured so that funds could be used to finance a hedge against drought.[3] In this way, a development lending operation can support assistance with risk management frameworks, strengthening technical skills, as well as strategy implementation.

If a government decides to use price risk management instruments to manage exposure to commodity price volatility, guidelines on restricting transactions to counterparties with sound financial standing should be established, along with an ongoing process for regularly monitoring ratings and identifying events that may lead to a deterioration of counterparty credit quality. In addition, counterparty credit limits must be in place to effectively manage the level of overall credit risk faced by the government as the buyer/user of a commodity derivative. Exposure limits measure the overall exposure for a particular counterparty, and threshold levels are trigger mechanisms that initiate collateral calls once the exposure level reaches the threshold level. Collateral agreements lay out guidelines for collateral calls, which can be used to lower effective exposure. These agreements can be drafted, based on the ISDA Master Derivatives Agreement and Credit Support Annex, currently used as the basis for most commodity hedge transactions.

After the hedges are executed, it is prudent to monitor whether they perform as designed. In some cases, external factors can alter the government's risk exposure, which could require adjustments in the hedged positions. Changes in the value of exposure over time can be detected using periodic, mark-to-market measurement (i.e., revaluation of the exposure using current market prices).

Successful execution of a hedging strategy thus depends on solid systems and robust operational procedures. Operational support for derivatives transactions should include trade capture, counterpart confirmation, documentation, settlement of cash flows, generation of accounting entries, account maintenance and reconciliation, valuation, reporting, risk, and compliance. Additional issues important to consider are (i) preparing and executing legal documentation to support derivative transactions; (ii) managing relationships with financial counterparts (investment banks, fiscal agents, clearing brokers, custodians, and other banking institutions) with respect to transaction support; (iii) providing settlement, accounting, custody, and compliance services in support of transactions; and (iv) generating post-transaction reports, risk, and performance (box 5.6).

Conclusion

Although the use of commodity risk management tools is not widespread among nations, the business of commodity hedging is well-established in the commercial sector, and hedging tools are used on a daily basis by commodity producers, consumers, merchants, financiers, trading companies, and brokerage firms. This chapter has presented an overview of the physical and financial hedging instruments available, described critical

Box 5.6

Technical Expertise Needed to Support Commodity Risk Management

When a commodity risk management strategy relies on financial instruments, execution of the strategy is done using market transactions that are carried out according to determined parameters, sequencing, and maturities. Supporting this process requires risk managers, accountants, information technology specialists, and internal auditors with knowledge of and expertise in risk management, financial transactions, and commodity market operations.

Generally speaking, technical support for these processes is divided into three independent offices, which would typically be housed in the finance ministries, treasury departments, or other agencies accustomed to conducting financial transactions on behalf of the government.

The three functional offices to support such operations are:

- *Front office*, with staff experienced in managing relationships with investment banks, using market-based information systems (e.g., Bloomberg or Reuters), transacting market operations, and using internal systems to appropriately record and monitor transactions.
- *Middle office*, with staff experienced in analyzing markets; quantifying risks; and preparing, reporting on, and monitoring strategy performance.
- *Back office*, with staff experienced in processing, accounting for, recording, and monitoring financial transactions.

Source: Authors.

steps in risk assessment, examined various approaches for technically evaluating hedging strategies, and discussed operational requirements (box 5.7).

Developing Guidance on Best Practices

Government ministries or power-sector entities inexperienced in commodity risk management may want to consider technical support and/or training to build capacity in these areas. In addition to covering the issues listed above, the components of such training might cover the following:

- Risk measurement using sensitivity analysis, scenario analysis, VaR, Monte Carlo, and other technical techniques.

Box 5.7

Summary Checklist for Institutions Implementing Commodity Risk Management

The fundamental steps supporting commodity risk management are well-established in the commercial world, and apply to any interested organization, whether public or private. The framework for implementing a commodity hedging strategy should always cover the following:

- Analytical work that assesses the impacts of short-term price volatility and carefully quantifies price exposure.
- Documentation of the objectives of the hedging strategy.
- Documentation of the reasons for selecting a specific hedging product, including details on terms of coverage.
- Description of operational arrangements, including roles and responsibilities of actors, agencies, and mechanisms for cost accounting and managing potential payouts associated with the hedging instrument.
- Verification of adequate legal and regulatory infrastructure to support the use of commodity derivatives; this involves ensuring that contracts are legally enforceable and not rendered invalid by local law, as well as establishing legal assurances that individuals who make decisions related to the transactions have the authority to do so.
- Establishment of procedures for those authorized to (i) negotiate, (ii) approve, (iii) execute, and (iv) audit transactions and reports; these authorizations should include the limits within which those individuals are authorized to act.
- Verification of issues related to negotiating ISDA Master Derivatives Agreements with market counterparties, with a specific focus on ensuring that equitable terms are reflected in the agreements.
- Review of accounting policies and tax regulations related to the management of hedging instruments.
- Establishment of procedures to select counterparties and brokers; this could involve establishing requests for proposals (RFPs) to help guide choice and evaluation of counterparties.
- Setting of clear limits for acceptable market and counterparty risks.
- Establishment of back-office and control procedures for monitoring and managing the hedge position, including the process for valuing transactions; this could include support to an operations department responsible for providing

(continued next page)

Box 5.7 *(continued)*

accounting, settlements, documentation, valuation, financial management reporting, and internal control services.

• Development of procedures for oversight and supervision of and reporting on risk management operations.

Source: Authors.

- Fundamentals of the energy risk management markets.
- Overview of financial risk management tools (e.g., futures, options, collars, and swaps).
- Basics of the ISDA Master Derivatives Agreement and Credit Support Annex.
- Back-office issues and control systems.
- Support for policy-related issues and establishment of institutional frameworks.

Since most governments in the region lack experience in these areas, it will be important over the next few years to document and explore good practice, both regionally and globally, in commodity risk management. In addition to the detailed operational aspects described above, a best practice guide should include information on what drives the decisions of policy makers, institutional reactions, how expertise is developed or acquired, and costs in terms of financial resources and time. It should also provide a more detailed description of how each of the above-discussed issues affects individual actors in the sector—including public and private generators, importers, distributors, and government entities—who manage various aspects of producer or consumer activity. As an example, a generator and government-owned distribution company with a strict tariff regime are affected differently by price volatility and, as a result, more detailed work in this area is needed to provide more information on how specific entities in the supply chain can use these tools. Development partners and countries will need to work together to share experiences and build this knowledge base, perhaps through regional workshops and a collaborative approach to developing a best practice guide.

The Role of the World Bank

For many years, the World Bank has assisted countries in developing sound practices in public debt management designed to strengthen financial resilience. Traditionally, member countries' focus has been on managing shocks related to either endogenous economic conditions or external financial indicators, such as interest and/or currency rates. As a result, countries have been encouraged to maintain domestic macroeconomic stability and attempt to hedge exposure to interest rate and currency fluctuations. During the recent global financial crisis, it became clear that many countries benefited from this work. Despite the scale of the crisis, to date it has not resulted in a sovereign debt crisis among emerging market countries. Two significant factors in this outcome are improved macroeconomic management and public debt management in these countries over the past decade (Anderson, Silva, and Velandia-Rubiano 2010).

Alongside efforts to strengthen capacity in public debt management, the World Bank has implemented changes in its financial product menu, which are designed to help facilitate flexibility in financing and access to risk management markets. In 1999, the Bank began to expand its menu of financial products and services in an attempt to bridge the gap between, on the one hand, evolving and progressively sophisticated international financial markets and products and, on the other, sovereign borrowers with limited access to and knowledge of such markets and products. Today, many middle-income market economies enjoy reasonable, although oftentimes sporadic, access to international financial markets and have become increasingly comfortable in the use of what have become standardized, widely-used financial products.

For commodity risk management, the World Bank offers member countries the ability to structure loans so that repayment obligations are linked to commodity price protection through a commodity swap (box 5.8). Countries interested in assistance in this area can also request that the Bank support hedging programs by acting as an advisor or executing agent, or, potentially, playing an intermediation role by acting as counterparty to the country while simultaneously entering into a back-to-back contract with a market counterparty. Depending on the needs, an advisory or execution agreement could include (i) technical work to support risk assessment and/or simulate the outcomes and costs/benefits of various hedging strategies, (ii) assistance linking the hedging strategy to the formal policy framework, (iii) technical capacity-building for staff on handling relationships with counterparties and evaluating pricing, and (iv) assistance in arranging all aspects of the execution strategy

Box 5.8

IBRD Commodity Hedges

The International Bank for Reconstruction and Development (IBRD) currently offers borrowers access to commodity swaps linked to new or existing IBRD loans. This product gives borrowers an opportunity to protect themselves from exposure to commodity prices by linking repayment obligations on IBRD loans to the commodity price. For an oil-importing country exposed to the risk of commodity price increases, the commodity swap can be structured so that repayment of the principal and/or interest rate on the IBRD loan would decrease if commodity prices increase. Conversely, for an oil-producing country exposed to the risk of commodity price declines, the commodity swap can be structured so that repayment of the principal and/or interest rate would decrease if commodity prices decrease.

An IBRD commodity hedge structured in this way would have two components:

- An existing IBRD loan, with a corresponding rate of interest based on the London Interbank Offered Rate (LIBOR) and pre-determined repayment characteristics.
- An overlying transaction that would exchange the cash flows of the original IBRD loan for a new set of cash flows, based on an interest rate and repayment profile that incorporate the costs and potential payouts of a commodity hedge, such as a swap. A commodity put or call option could also be the overlying commodity derivative that would establish the desired price protection.

Borrowers evaluating this instrument would need to consider (i) how much of the country's exposure to commodity price volatility to cover, (ii) which price levels to protect, and (iii) the tenor or time frame of the coverage.

Source: Authors.

and its implementation on behalf of the client (without acting as counterparty) (appendix B).

In addition to strengthening capacity in this area, the World Bank's convening power can be used to bring sovereign members with similar problems together to develop customized solutions and support efforts to approach the market in a coherent and comprehensive manner. As in other areas, the Bank can continue to play a role as a center for excellence, innovation, and international best practices. This is particularly important for middle-income countries owing to growing sophistication of and ability to access markets. For low-income countries, the focus on risk assessment and strengthening institutional frameworks may be more of an immediate priority, as these countries will continue to develop financial strength and the ability to use market approaches over time. Overall, it is important to recognize that the task of designing appropriate solutions relevant to developing countries involves a high degree of customization.

At a global level, the World Bank has been advocating for greater investment in risk management for many years, an effort that has been supported by the Sustainable Development Networks' Agricultural Risk Management Team, support to the UN High Level Task Force on the Global Food Security Crisis (2008), the Global Food Crisis Response Program (2009), the Global Agriculture Food Security Program (2010), and Treasury Advisory Services. Each of these programs offers countries technical support for (i) quantifying exposure to price risks, (ii) assessing the technical and commercial feasibility of transferring risks to the market, (iii) assisting with hedging product design, and (iv) building capacity to support implementation of risk management strategies.

In June 2011, in coordination with international partners, risk management was incorporated into the G20 Action Plan on Food Price Volatility and Agriculture, where it was recognized that financial instruments have an important role to play alongside critical investments in risk reduction, risk mitigation, production, and, in the case of commodities, market development. These same issues have relevance to energy price risk and efforts to strengthen energy security.

Notes

1. The term *hedging* should not be confused with *speculation*, which refers to the use of price risk management instruments for the purpose of profiting on short- or long-term price movements, independent of a direct interest in use of the physical commodity.

2. Non-price risks faced by distribution utilities (i.e., related to operational issues that increase the cost of technical and nontechnical losses) are a key element in the risk assessment. Since these issues can be related to tariff pass-through regulations, they can adversely affect any strategy's overall success. It should be noted that distribution utilities that have difficulty

managing technical and nontechnical losses may also have problems designing and implementing an efficient hedging strategy.

3. The Malawi Agricultural Development Program Support Project.

References

Anderson, P. R. D., A. C. Silva, and A. Velandia-Rubiano. 2010. "Public Debt Management in Emerging Market Economies: Has This Time Been Different?" World Bank Policy Research Working Paper 5399, World Bank, Washington, DC.

Gilbert, C. 2010. "Electricity Hedging." Background research prepared for the study, Mitigating Vulnerability to High and Volatile Oil Prices in Latin America and the Caribbean, World Bank, Washington, DC.

CHAPTER 6

Diversifying from Oil-Fired Power Generation

Reducing oil consumption by diversifying from oil-fired power generation can fundamentally manage the impact of high and volatile oil prices
on the power sector. But developing alternative energy sources to achieve
a diversified generation matrix requires time, which is a challenge for
many developing countries that struggle to provide sufficient capacity. In
many countries of Central America and the Caribbean, for example, the
share of oil-fired capacity has been growing in recent decades.

In Latin America and the Caribbean (LAC) and other developing
regions worldwide, climate-change concerns are spurring the development of renewable energy. Along with energy diversification, its benefits
include local resource use and cleaner energy production. In addition, it
helps to optimize the energy-generation portfolio since its cost is not correlated with oil prices, which could constitute up to 90 percent of the
operating costs of certain generation technologies (e.g., a combustion
turbine plant using distillates). The cost of electricity generated from
other fossil fuels, such as natural gas and coal, is correlated somewhat
with oil prices, but much less so than in the past.

Taken together, these benefits reduce overall volatility. This conclusion
is supported by recent studies that borrow from the portfolio models of
the finance literature (e.g., the mean-variance frontier and real options)

to determine and quantify the value from the optimal energy-generation system (Bazilian and Roques 2008). These models emphasize the price, quantity, and duration effects of energy disruptions in a non-diversified, energy-generation system (box 6.1).

The LAC region has a wide array of renewable resources and technologies available for diversifying its electricity portfolio. These range from wind in Argentina to hydroelectricity and biomass in Brazil (Yépez-García, Johnson, and Andrés 2011), to geothermal in Central America. In 2007, renewable energy represented about 59 percent of the region's total power generation—higher than in any other world region. Hydropower alone accounted for 57 percent of total generation, and is still considered the renewable option with the largest generation potential over the next two decades.

This chapter considers the potential for the LAC region to increase non-oil electricity generation through the use of renewable-energy sources and non-oil conventional thermal power.[1] Central America and the Caribbean provide basic examples of the potential for renewable

---

Box 6.1

Applying Portfolio Theory to Optimize Energy Generation

A key element of modern portfolio theory especially relevant to power-sector planners is correlated risk. Since highly correlated asset categories, such as energy prices, tend to move together, diversifying from one asset to another will not reduce overall portfolio risk. This insight has important implications for power-generation efforts that substitute one hydrocarbon fuel cycle for another in the name of energy diversification.

In applying portfolio theory to electricity planning, Shimon Awerbuch, the late financial economist, developed a method that analyzes the risk-return relationship to distinguish feasible portfolios from those that are either infeasible or undesirable. A feasible portfolio is said to meet the criteria for riskiness; that is, the risk of a given set of power-generation plants, expressed as the standard deviation of the portfolio returns, is at or below some target level. Generation portfolios that offer the same risk level with a lower return/higher cost are feasible, yet undesirable; and those that offer a risk level with unattainable returns are infeasible.

Awerbuch's analysis shows that the expected cost of a portfolio varies inversely with risk; that is, the greater the risk, the lower the total expected cost. Risk varies

(continued next page)

Box 6.1   *(continued)*

directly with portfolio return and inversely with unit generation cost. When the expected portfolio risk and return are plotted together, they form a "frontier," which separates feasible portfolios from infeasible ones, as shown below (box figure 6.1.1).

Portfolios "northwest" of the frontier are more desirable but less feasible, while those "southeast" of the frontier might be feasible but are less desirable since some other combination of power-plant investments can always generate power for a better mix of risk and return.

Box Figure 6.1.1   Risk-and-Return Frontier

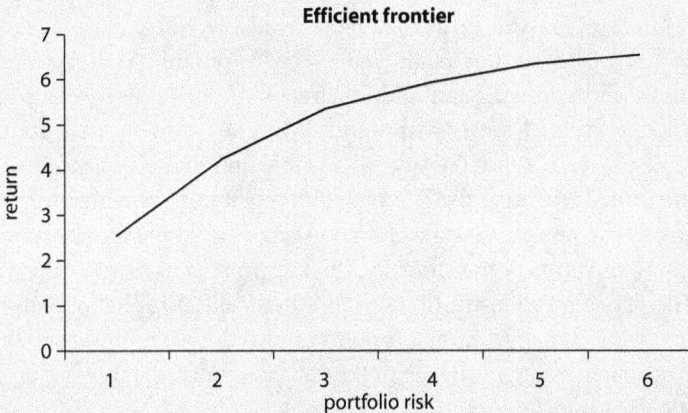

Sources: Awerbuch 2000; Awerbuch and Berger 2003.

generation, demonstrating how highly vulnerable countries can manage high and volatile oil prices to mitigate such risk (appendix C). The chapter then attempts to quantify and illustrate the potential benefits to economies, with a fuller discussion provided in chapter 9.

Potential for Non-Oil Generation: Central America and the Caribbean

To diversify their generation mix from oil-fired power, the subregions of Central America and the Caribbean could pursue three groups of alternatives: (i) hydropower, (ii) non-hydro renewable power (geothermal, biomass, wind, and solar), and (iii) non-oil conventional thermal power (natural gas and coal).

Estimates of the potential for renewable alternatives vary widely between countries and resources in terms of type, detail, and quality. In general, hydropower resources are the best understood, having been the most widely used for decades. However, in some countries, publicly available estimates are limited to technical potential. In terms of non-hydro renewable power, geothermal resources have been exploited on a substantial scale, especially in Central America. But the cost of developing credible estimates of potential is higher than for hydropower with less information in the public domain (Poole 2009).

Additional non-hydro renewable sources include biomass, wind, and solar photovoltaics (PV). In terms of biomass, sugarcane residue is a traditional source of on-site power, and the available resource is well documented. In this case, the main issues center on mill power-generation technology and configuration and incentives to encourage greater investment in exports of power to the grid. Wind, a relatively new source of renewable power, was not considered a viable option by most power-sector planners until quite recently. The type and quality of available information vary greatly by country (Poole 2009). Again, the key issues concern the technology and costs of transforming the known resources into energy, with the difficulty of more institutionally and politically complex operations.

In assessing the approximate potential of renewable-energy resources, three comparable categories of potential have been used to incorporate information from the literature, as follows:[2]

- *Technical potential.* This term refers to the gross potential identified that may meet some minimum cost-related criteria (e.g., the wind potential of sites with wind speeds above 7 m/s); these estimates tend to be generic and not based on the detailed analysis of individual sites.
- *Usable potential.* Adopted from Nexant (2010a), this is a more restrictive definition based on more detailed information from an inventory of projects, where emphasis is on the possible restrictions on development.
- *Effective potential.* This term, in principle, refers to the long-term market potential of implementable projects; the ratio of effective to usable potential is usually substantially higher than to technical potential.

The subsections that follow review the available information on the potential of hydropower and non-oil conventional thermal power (natural gas and coal) to diversify the generation mix from oil in Central America and the Caribbean. A description of the potential for non-hydro renewable power—ranging from geothermal and biomass (sugarcane bagasse) to wind and solar energy—is provided in Poole (2009).

Hydropower

Hydroelectric resources are unevenly distributed among countries in Central America and the Caribbean. As table 6.1 illustrates, the more refined estimates of the usable potential that could ultimately be implemented are substantially smaller than the estimated technical potential.

In most Caribbean island nations, existing hydropower output is quite small and the remaining potential either insignificant or non-existent. A partial exception is the Dominican Republic; but with the inventoried

Table 6.1   Summary of Remaining Hydropower Potential

Country	Output (2007) (GWh)	Technical potential MW	Technical potential GWh	Usable potential MW	Usable potential GWh
Costa Rica	6,770			6,633	29,123
Grenada	0				
El Salvador	1,740			2,165	9,349
Guatemala	3,010	10,890		5,784	21,443
Haiti	480				
Panama	3,870			3,040	13,328
St. Lucia	0				
Barbados	0				
Dominica	0				
Dominican Republic	1,680			535	1,541
Guyana	0	7,600	19,600		
Honduras	2,300			4,991	21,861
Jamaica	170				
Nicaragua	310			2,400	10,267
St. Vincent and the Grenadines	0				
St. Kitts and Nevis	0				
Belize	90				
Antigua and Barbuda	0				
Suriname	1,360	2,208	9,222		

Sources: OLADE 2008; Poole 2009; Nexant 2010a for Central America potential; CNE-DR 2006 for Dominican Republic potential; OLADE 2005 for Guatemala technical potential; OLADE 2008 for other technical potential.
Note: Shaded cells indicate that potential is insignificant or estimates are unavailable.

potential, it would be difficult for that country to maintain its existing hydro share in the generation mix (estimated at about 11 percent) for even a few years.

The situation differs markedly in the Guyanas. Guyana has a technical potential of 7.6 GW or nearly 20 TWh, far larger than its current consumption. None of this potential has yet been exploited. The 154-MW Amaila Falls Project represents the first major step to shift toward hydropower. However, the project raises many social and environmental issues, and experience with its approval and implementation may strongly affect prospects for further development of the country's hydro potential. Suriname is already supplied mainly by hydro, which represents 84 percent of total generation. In principle, it should be possible to maintain the high share of hydro over the coming decade, but interconnection and exchange with Guyana may be helpful.

In Central America, the remaining potential is large enough, in theory, to permit an increase in hydro's share of regional generation. However, the inventory of projects being actively evaluated for expansion planning is much smaller, at 7,000–8,500 MW.[3] The per-kilowatt investment and socio-environmental impacts of hydropower projects vary widely by site. Beyond the need for economic viability, projects must have acceptable social and environmental impacts; however, it is unclear whether the inventory presented by OLADE (2008) took these considerations into account.

In addition, the viability of a number of larger projects may require greater regional integration. Investment needs for a large hydro scenario are comparatively high and may be difficult to mobilize, especially if the sector is struggling to keep pace with demand growth. Taking these factors into account, it would be an achievement to simply maintain Central America's current share of hydro generation after two decades of steady decline.

Non-Oil Conventional Thermal Power

In addition to renewable energy sources, Central America and the Caribbean have other alternative choices for diversifying their power-generation matrix. Natural gas and coal could help to reduce oil dependency and attenuate the impact of oil prices in these two subregions. Even though natural gas prices in North America are correlated with oil prices, such correlation has decreased in recent years as a result of new gas discoveries in the region (figure 6.1). A lower correlation between oil and natural gas prices definitively helps to diversify the power-generation

Figure 6.1   Historical Evolution of Oil and Natural Gas Prices, 2000–12

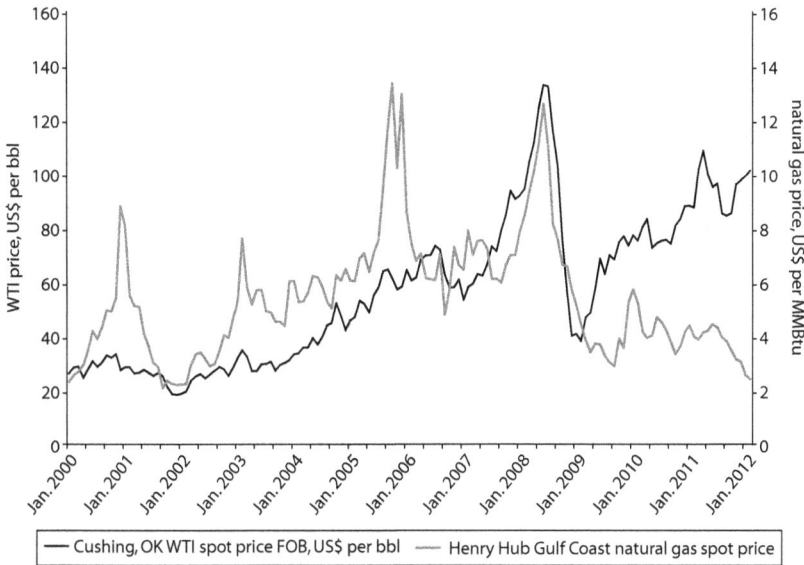

Source: Authors, with EIA data (www.eia.gov).

portfolio and reduce oil price vulnerability. In addition, coal may further reduce vulnerability as coal and oil prices have a low correlation. We discuss these two options below as additional alternatives to oil-fired power generation.

Natural gas. Among all countries in the two subregions, Trinidad and Tobago is the only significant producer of natural gas, and the Dominican Republic has some generation using liquefied natural gas (LNG). Central America has no generation from natural gas. Thus, for the vast majority of countries, expanding natural gas use for electricity generation would mean importing the fuel.

The dramatic expansion of unconventional gas production in North America has radically changed the supply-demand balance. Only five years ago, domestic gas supplies were tight, and the United States was viewed as an important future market for imported LNG. It now appears that the unconventional gas supply will continue to expand, and several projects to export LNG are under way. Since 2009, North American natural gas prices have completely decoupled from the price of oil (figure 6.1).

These new developments suggest the potential for LNG as a viable short- and medium-term option to help countries in the Caribbean sub-region to diversify generation away from oil and thus reduce vulnerability to high and volatile oil prices. The countries with the greatest potential to initiate LNG consumption are Haiti, Jamaica, and Barbados. Given its existing LNG import facility, the Dominican Republic is well-positioned to expand consumption. Likewise, countries in Central America could help to meet growing demand using natural gas in the form of either pipeline gas or LNG.

However, a natural gas import strategy is not without constraints. Natural gas import projects require substantial investments in pipelines and LNG receiving terminals, tankers, and other infrastructure, the cost of which must be amortized over many years and recovered in end-user prices. Also, gas supply contracts normally include substantial take-or-pay obligations covering 80 percent or more of the contracted volume. As a result, the commercial structure of import projects can be highly complex, and the credit capacity of buyers a key limitation. In addition, competition for long-term LNG supply is intense, and most LNG is traded at prices that, unlike in the United States, are closely tied to those of oil or petroleum products. Until supply increases—from the United States or elsewhere—buyers may find that natural gas does not generate substantial cost savings compared to oil. That said, for creditworthy buyers able to aggregate markets of sufficient size to realize economies of scale, natural gas can bring about important diversification in fuel supply.

Coal. The use of coal for electricity generation is insignificant in Central America and the Caribbean, limited to only a 314-MW plant in the Dominican Republic. However, coal-fired units have been projected in a number of power-sector expansion plans, especially in Central America. Some capacity has already been contracted, and up to 2,000 MW are being considered for regional expansion plans in Central America during 2010–22.

Shifting to coal is a conventional strategy to diversify from oil in electricity markets above a certain size. However, it is unclear whether coal offers cost advantages when compared to natural-gas units of similar size. In addition, the carbon footprint of coal-fired systems is substantially larger. Although new clean-coal technology can reduce carbon dioxide (CO_2) emissions per kilowatt hour, emissions are still significantly higher than for natural gas. Such plants are also likely to have higher capital costs than the basic designs considered to date.

The expansion of conventional thermal-generation technologies using fossil fuels other than oil seems inevitable in Central America and the Caribbean. Renewable resources alone cannot provide all of the needed expansion of electricity supply at an acceptable cost. However, compared with natural gas, coal may not be the cheapest solution from the perspective of the grid—even before considering the option's impact on greenhouse gas (GHG) emissions.

Potential Avoided Impact from Oil-Fired Generation

Today renewables account for about 57 percent of electricity generation in Central America and 14 percent in the Caribbean and Guyanas. The difference results mainly from the larger share of hydro in Central America's energy mix, and to a lesser extent, the higher share of non-hydro renewables (12 percent in Central America versus none in the Caribbean). Nearly all of the remaining generation is oil-based, with the exception of the Dominican Republic, where natural gas, as well as coal, contributes to the energy supply, and Guatemala, which utilizes some coal (table 6.2).

To identify the renewable-energy potential in a specific region or country, several factors must be considered, and uncertainty affects the estimates for the various sources. In this context, tables 6.3 and 6.4 incorporate information from diverse sources, including the tables in this chapter and Poole (2009), to approximate a picture of the potential contributions of renewable resources to system expansion and diversification from oil in the two subregions.

Table 6.3 shows estimates for four of the renewable resources considered. In the case of hydro, geothermal, and wind, estimates are for potential that may ultimately be developed over the next several decades. Assumptions are made about the share of technical and usable potential that may ultimately be exploited. For example, in the case of hydro, it is assumed that 70 percent of usable potential, versus 30 percent of technical potential, could ultimately be developed (note, table 6.3). These cases have no particular time horizon for development.

In the case of sugarcane, it is assumed that 65 percent of the usable potential could be developed by about 2030. In this case, the underlying energy resource is a flow of residues whose volume changes, depending on such factors as economic activity. A similar situation exists for urban solid waste and residues from sawmills and other industries, which are not covered in this report; this situation is quite distinct from hydro, geothermal, and wind, whose underlying potential is defined by natural geography.

Table 6.2   Share of Energy Sources in the Generation Mix of Central America and the Caribbean, 2007

Country	Hydropower (%)	Other renewable (%)	Oil (%)	Coal and gas (%)
Costa Rica	74	18	8	0
Grenada	0	0	100	0
El Salvador	30	26	44	0
Guatemala	34	10	44	12
Haiti	84	0	16	0
Panama	60	0	40	0
St. Lucia	0	0	100	0
Barbados	0	0	100	0
Dominica	0	0	100	0
Dominican Republic	11	0	67	22
Guyana	0	0	100	0
Honduras	36	6	58	0
Jamaica	2	0	98	0
Nicaragua	10	15	76	0
St. Vincent and the Grenadines	0	0	100	0
St. Kitts and Nevis	0	0	100	0
Belize	43	8	49	0
Antigua and Barbuda	0	0	100	0
Suriname	84	0	16	0
Central America	45	12	40	3
Guyanas	55	0	45	0
Caribbean	10	0	78	13

Sources: OLADE (2008) and Nexant (2010b) for smaller Caribbean islands, except Grenada and Barbados.

One can see that hydro is by far the largest renewable energy source overall. Non-hydro renewables also figure prominently. In Central America, they amount to 44 percent of the hydro resource, while in the Caribbean and Guyanas, their potential is substantially larger than that of hydro.

Table 6.4 takes the totals for hydropower and other renewables and compares them with generation in 2008 (both total and oil-based). The enormous differences between countries and their endowments are striking. Many have effectively exploitable renewable resources that are multiples of total current generation (excluding generation from oil). In such cases, it should be possible to supply a large share of system expansion with renewables, while steadily backing away from oil-based generation, assuming an appropriate policy context.

As a reference, if electricity demand increased by 4.5 percent annually to 2030, growth over the period would amount to 175 percent.

Table 6.3   Remaining Effective Potential of Renewable Resources

Country	Hydropower (GWh)	Geothermal (GWh)	Wind (GWh)	Sugarcane[a] (GWh)	Non-hydro renewables (GWh)
Costa Rica	20,386	1,621	1,127	293	3,041
Grenada	0	0	0	0	0
El Salvador	6,544	2,605	*2,798*	273	5,676
Guatemala	15,010	3,630	*2,124*	1,075	6,830
Haiti	0	0	*4,684*	66	4,751
Panama	9,329	206	*2,139*	128	2,474
St. Lucia	0	123	75	0	198
Barbados	0	0	19	27	46
Dominica	0	491	0	0	491
Dominican Republic	1,079	0	*5,897*	380	6,277
Guyana	*5,880*	0	0	239	239
Honduras	15,302	545	*2,883*	320	3,747
Jamaica	0	0	132	134	265
Nicaragua	7,187	5,377	*6,014*	147	11,538
St. Vincent and the Grenadines	0	0	4	1	5
St. Kitts and Nevis	0	1,471	9	11	1,492
Belize	0	0	0	80	80
Antigua and Barbuda	0	0	180	0	180
Suriname	*2,767*	0	0	6	6
Central America	73,759	13,983	17,086	2,236	33,305
Guyanas	8,647	0	0	245	245
Caribbean	1,079	2,085	11,000	700	13,785

Source: Poole 2009.
Note: Values in italics are based on estimates of technical potential; all others are based on usable potential.
a. The effective potential from sugarcane is based on the low value for the 65-bar steam system, using 50 percent of field residues, shown in Poole (2009); in Central America, existing generation sold to the grid has been subtracted from the estimated potential.

Thus, in theory, renewable resources would provide for all system expansion to 2030 in Central America, Guyanas, and several Caribbean islands. Of course, this cannot and should not occur for many reasons; yet a prima facie argument can be made that a substantial share of expansion—60–70 percent in the majority of countries and an even higher percentage in some—could be based on renewable resources. Conversely, for a group of Caribbean countries, the renewable resources considered are relatively small compared to the expansion need.

Another perspective focuses more specifically on oil-fired generation capacity and the possibilities for its substitution. Table 6.5 illustrates the impact of 10 percent of effective renewable energy potential on

Table 6.4   Effective Potential versus Generation of Renewable Resources in 2008

Country	Total generation (GWh)	Hydropower (GWh)	Non-hydro renewables (GWh)	Total renewables (GWh)	Renewable versus 2008 generation (%)
Costa Rica	9,484	20,386	3,041	23,427	247
Grenada	171	0	0	0	0
El Salvador	5,639	6,544	5,676	12,220	217
Guatemala	8,717	15,010	6,830	21,840	251
Haiti	780	0	4,751	4,751	609
Panama	6,427	9,329	2,474	11,803	184
St. Lucia	331	0	198	198	60
Barbados	1,023	0	46	46	4
Dominica	87	0	491	491	564
Dominican Republic	15,415	1,079	6,277	7,356	48
Guyana	868	5,880	239	6,119	705
Honduras	6,536	15,302	3,747	19,050	291
Jamaica	3,962	0	265	265	7
Nicaragua	3,360	7,187	11,538	18,724	557
St. Vincent and the Grenadines	132	0	5	5	4
St. Kitts and Nevis	195	0	1,492	1,492	765
Belize	212	0	80	80	38
Antigua and Barbuda	318	0	180	180	57
Suriname	1,619	2,767	6	2,773	171
Central America	40,163	73,759	33,305	107,064	267
Guyanas	2,487	8,647	245	8,892	358
Caribbean	22,626	1,079	13,785	14,864	66

Sources: Previous tables in this chapter; Poole 2009.

substitution of oil-fired capacity in 2007. It compares the scale of the renewable resource potential with existing oil-fired output in the countries of Central America and the Caribbean, showing how much oil-fired generation would be reduced if substituted by 10 percent of the effective renewable potential.

Even with this small share of renewable-energy potential, the estimated substitution in some countries is larger than existing oil-fired output. In Costa Rica and Suriname, the reason is that the current share of electricity generated by oil-fired plants is small. In Haiti, however, the public electricity supply is disproportionately small for a country of its size; this was so prior to the 2010 earthquake. While there are no statistics on self-generation, it is probably substantially larger than the public

Table 6.5   Scale Comparison of Renewable Resources to Current Oil-Fired Generation

Country	Oil-fired capacity in 2007			Impact of 10% of effective renewable-energy potential on substitution of oil-fired capacity		
	Share of generation (%)	Output (GWh)	Consumption (10^3 boe)	Output (GWh)	Consumption (10^3 boe)	Oil-fired generation (%)
Costa Rica	8	727	961	2,343	3,097	322
Grenada	100	170	219	0	0	0
El Salvador	44	2,564	3,747	1,222	1,786	48
Guatemala	44	3,809	4,118	2,184	2,361	57
Haiti	16	90	559	475	2,951	528
Panama	40	2,600	3,958	1,180	1,797	45
St. Lucia	100	331	622	20	37	6
Barbados	100	950	1,899	5	9	0
Dominica	100	87	103	49	58	56
Dominican Republic	67	9,966	15,467	736	1,142	7
Guyana	100	870	1,419	612	998	70
Honduras	58	3,661	5,793	1,905	3,014	52
Jamaica	98	7,310	5,705	27	21	0
Nicaragua	76	2,425	3,763	1,872	2,906	77
St. Vincent and the Grenadines	100	132	204	1	1	0
St. Kitts and Nevis	100	195	366	149	280	77
Belize	49	104	195	8	15	8
Antigua and Barbuda	100	318	597	18	34	6
Suriname	16	260	404	277	430	107
Central America	40	15,786	22,340	10,706	14,960	68
Guyanas	45	1,130	1,823	889	1,428	79
Caribbean	78	19,653	25,936	1,486	4,547	8

Sources: OLADE (2008) and Nexant (2010b) for consumption and output of oil-fired capacity; previous tables in this chapter and Poole (2009) for renewable energy potential.

supply. Even so, renewable energy in the form of wind power could play a significant role.[4]

Beyond these special cases, one may observe a division between the Central American countries and the Guyanas on the one hand and most of the Caribbean islands on the other. For the former group, a 10 percent share of effective renewables potential could substitute for about 50–70 percent of existing oil-fired output. However, for the group of Caribbean countries, the average is only about 8 percent, while for some the calculated potential is zero. These countries are also among those that depend most heavily on oil-based generation with the highest supply costs.

Some of the larger countries could mitigate the situation via natural gas imports. The Dominican Republic has already done so, and Jamaica and Barbados are likely to follow. Such countries as Dominica and St. Kitts and Nevis have substantial geothermal reserves, which, if developed under relatively large projects, could transform their power supply mix. However, in the other countries, the scope for substitution of oil-fired capacity appears quite limited. As a note of caution, renewable-resource estimates for many Caribbean islands are still preliminary. Thus, the effective renewable-energy potential among the resources quantified may be larger, in part, because the baseline cost of electricity supply will remain higher.

As discussed in Poole (2009), photovoltaic (PV) systems are already close to competing with diesel-based generation. Hybrid diesel-PV systems could soon offer the most economical solution for small island markets. If PV costs continue to decline, which appears likely, the attractiveness of this technology, usually in on-site applications, could rapidly expand beyond this niche market.

The estimates of potential discussed in this chapter are necessarily preliminary. Countries throughout Central America and the Caribbean have already begun to assign renewable resources increased weight in their near-term expansion plans. As more experience is gained, the quality of the estimated effective potential of these various resources should likewise improve substantially.

Summary Remarks

The analysis in this chapter has illustrated the potential of renewable energy to comprise a greater share of power generation in both Central America and the Caribbean and thus reduce medium- and longer-term vulnerability to high and volatile oil prices. Among the renewables considered, biomass could offer immediate output gains as long as appropriate retrofitting is put in place. From a policy perspective, geothermal has a large potential to diversify the power system, although exploration costs remain a barrier to resource exploitation (World Bank/ESMAP 2009, 2011).

In combination with fossil-fuel alternatives, especially natural gas, increasing the share of renewable-energy sources in Central America and the Caribbean can yield energy-security, economic, and environmental benefits. Together with greater fuel efficiency, discussed in the next chapter, and an increasingly integrated regional power market, the subject of chapter 8, significant gains in energy security, as well as a reduction in GHG emissions, can be achieved.

Notes

1. The portfolio analysis and optimization method used in this report is based on the approach developed by Shimon Awerbuch, which calculates the risk and return (generation cost) of any portfolio of generation assets (box 6.1).
2. Unfortunately, the criteria of published renewable-potential estimates are seldom made clear; judgment was used in the review of the literature.
3. A list of CEAC candidate projects totals 6,900 MW (CEAC 2007), while projects being considered in national expansion plans total 7,000 MW (World Bank/ESMAP 2009); however, estimates for each country differ considerably. If one sums the higher estimate for each country, the regional total would be about 8,650 MW.
4. In Haiti, the highest-priority need would be to extend the public electricity grid, whose growth could be quite high; interestingly, wind power potential is dispersed in small areas throughout the country.

References

Awerbuch, S. 2000. "Getting It Right: The Real Cost Impacts of a Renewables Portfolio Standard." *Public Utilities Fortnightly*, February 15.

Awerbuch, S., and M. Berger. 2003. "Applying Portfolio Theory to EU Electricity Planning and Policy Making." IEA/EET Working Paper, International Energy Agency, Paris.

Bazilian, M., and F. Roques, eds. 2008. *Analytical Methods for Energy Diversity and Security. Portfolio Optimization in the Energy Sector: A Tribute to the Work of Dr. Shimon Awerbuch*. Amsterdam: Elsevier.

CEAC (Central American Electrification Council). 2007. "Plano Indicativo Regional de Expansion de la Generación Periodo 2007–2020." Grupo de Trabajo de Planificación Indicativa Regional (GTPIR), Consejo de Electrificación de America Central, Tegucigalpa.

Nexant. 2010a. *Promoting Sustainable Energy Integration in Central America*. Assessment for USAID El Salvador and USAID Central America and Mexico Regional Program. San Francisco, CA: Nexant.

————. 2010b. *Caribbean Regional Electricity Generation, Interconnection and Fuels Supply Strategy: Interim Report.* Prepared for the World Bank (January). San Francisco, CA: Nexant.

OLADE (Latin American Energy Organization). 2005. *Prospectiva Energética de América Latina y el Caribe.* Quito: Organización Latinoamericana de Energía.

————. 2008. *Energy Statistics Report—2007.* Quito: Organización Latinoamericana de Energía.

Poole, A. D. 2009. "The Potential of Renewable Energy Resources for Electricity Generation in Latin America." Working paper prepared for the report, Latin America and Caribbean Region's Electricity Challenge, World Bank, Washington, DC.

World Bank/ESMAP (Energy Sector Management Assistance Program). 2009. *Central America Sector Overview: Regional Programmatic Study for the Energy Sector—General Issues and Options Module.* Washington, DC: World Bank.

————. 2011. *Drilling Down on Geothermal Potential: A Roadmap for Central America.* Washington, DC: World Bank.

Yépez-García, R. A., T. M. Johnson, and L. A. Andrés. 2011. *Meeting the Balance of Electricity Supply and Demand in Latin America and the Caribbean.* Directions in Development. Washington, DC: World Bank.

CHAPTER 7

Investing in Energy Efficiency

Energy efficiency lowers electricity consumption and power-generation requirements, thereby reducing the need for imported oil and oil-derived products. Like renewable energy–based electricity generation, energy-efficiency measures require upfront investments, whose costs are recovered through the ensuing energy savings. Unlike power generation, however, the reach of energy-efficiency investments extends beyond power-sector generation, transmission, and distribution to encompass the industrial, commercial, and residential sectors. It is recommended that energy-efficiency measures be incorporated into both long- and short-term strategies aimed at reducing vulnerability to oil price fluctuations and market trends, owing to their fuel-saving nature and ability to avoid investments in new generation capacity.[1]

This chapter considers the energy-efficiency measures available to countries in Central America and the Caribbean to reduce fuel consumption and thus vulnerability to high and volatile oil prices. The next section discusses supply-side opportunities in the power sector; technical and commercial losses are analyzed to estimate the fuel savings from avoided combustion of oil derivatives and options for reducing these losses in country-specific contexts. The chapter then turns to demand-side opportunities to increase energy efficiency in the industrial, commercial, and residential sectors. This is followed by an attempt to quantify the potential savings from greater energy efficiency in oil consumption for electricity generation. Finally, general policy recommendations for overcoming barriers to energy efficiency are presented.

Supply-Side Efficiency

The following subsections describe the opportunities available to improve the supply-side efficiency of electricity systems in transmission and distribution grids. For each of the countries studied, energy losses are identified and the impacts of energy-loss reductions on oil price vulnerability are estimated. Also highlighted are some of the challenges to reducing losses and suggested general policy measures.

Assessing Technical and Commercial Losses

Transmission and distribution losses are grouped into technical and nontechnical or commercial losses. Reducing technical losses contributes to improving overall system efficiency and thus reducing fuel consumption; thus, it is considered an instrument that directly mitigates exposure to oil price volatility.

In Central America and the Caribbean, it is difficult to assess technical losses and thus the potential to reduce them since only aggregate-loss data are available for most countries.[2] Building on an analysis of available disaggregated data, along with other countries' data, this study approximated the breakdown of technical losses (TL) and commercial losses (CL) in 16 countries for which reliable data was available.[3] The difference between actual and efficient losses yields the scope for potential loss reductions for each country. Potential energy savings were calculated as the sum of full TL and 30 percent CL reductions. It was found that nearly all countries could reduce losses and thus avoid some generation (figure 7.1).

A further analysis was conducted to calculate the savings in diesel and heavy fuel oil (HFO) from fully harnessing the potential loss reductions in each country. This analysis is based on the characteristics of each country's marginal generation; it uses a static model with instantaneous effects of loss reduction in energy and fuel savings, while all other conditions remain constant. As figure 7.2 shows, Honduras, Jamaica, the Dominican Republic, Nicaragua, Panama and Guatemala would benefit the most from replacing what are presumably small, obsolete generating plants.

The analysis shows potential fuel savings by achieving efficient levels of losses in all countries. Diesel fuel may be saved in 15 countries, with

Figure 7.1   Energy Savings Potential in Selected Countries of Central America and the Caribbean

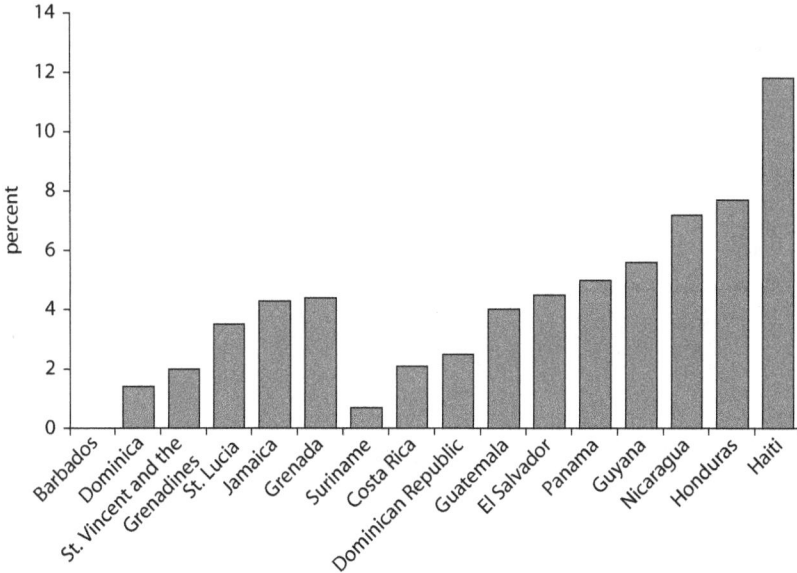

Source: Aburto 2010.

the largest savings accrued in the Dominican Republic and Jamaica, followed by El Salvador. Four countries—Guatemala, Honduras, Nicaragua, and Panama—eliminate the use of diesel fuel entirely and still achieve additional HFO savings.

Electrical losses are an inevitable consequence of energy flows through electricity transmission and distribution grids. The level of losses is a function of various factors, including grid configuration and design characteristics, degree of obsolescence, demand profile and composition, and operating practices. Therefore, it is difficult to identify the optimal level of losses for the particular grids and compare performance among them. Utilities should seek to minimize TL and eliminate the causes of CL to avoid excess losses that could seriously affect their financial health.

Options to Reduce Technical Losses

Variable. A grid's variable losses are approximately proportionate to the square of the current. Therefore, greater utilization of its capacity adversely affects losses. By increasing the cross-sectional area of lines and

Figure 7.2   Potential Savings in Fuel Oil Products from Supply-Side Loss Reduction in Selected Countries

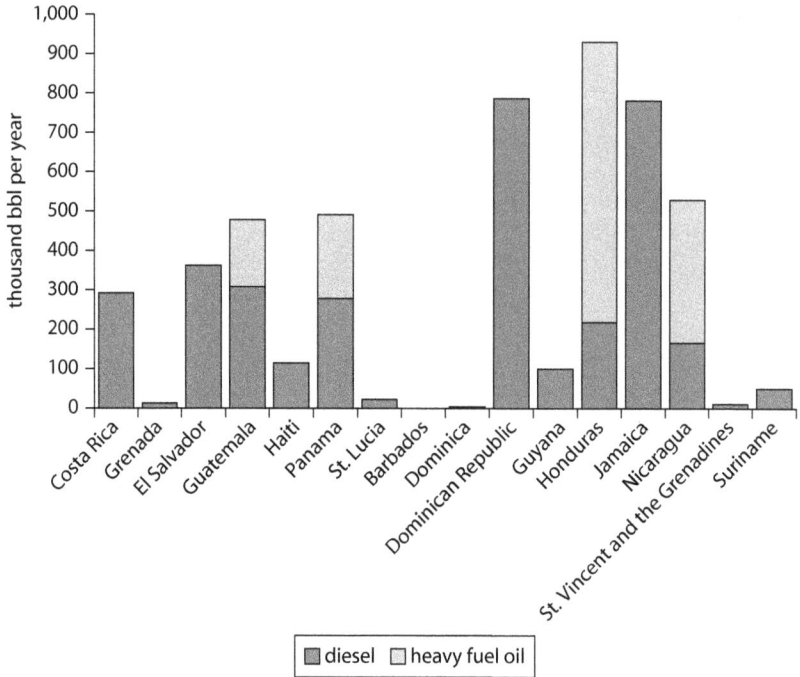

Source: Aburto 2010.

cables for a given load, losses decrease, which leads to a direct trade-off between the cost of losses and capital expenditure.

Fixed. There are five major ways to reduce fixed losses: (i) choose transformer technology carefully, (ii) eliminate transformation levels, (iii) switch off transformers, (iv) improve low-power factors, and (v) distribute generation. The level of fixed losses in a transformer depends, in large part, on the quantity and quality of the raw materials in the core. Transformers with more expensive core materials, such as special steel or amorphous iron cores, incur lower losses. Thus, in selecting transformers, there is a direct trade-off between capital expenditure and cost of losses.

A second way to reduce fixed losses is to eliminate transformation levels on the grid. Although some offsetting increase in variable losses may occur on a specific voltage grid, it often will be offset by the reduction in fixed losses. A third method is to switch off transformers during low-demand periods. Assuming a substation requires two transformers of a certain size during peak periods, one could be switched off during a low-demand period to reduce fixed losses. This would produce some offsetting increase in variable losses and might affect security and quality of supply, as well as the operational condition of the transformer. These trade-offs should be examined and optimized before reaching a decision about operating practices.

Higher losses are also caused by reactive power flows; these reduce the grid's effective capacity, which, as a share of installed capacity, is referred to as the power factor. Electrical motors and other apparatuses create reactive power and thus lower the grid's power factor. Consumers can improve their own load power factor by installing compensation equipment, but this will not occur unless they are charged for the low power factor. Such charges should be identified on customer bills.

Finally, locating generation closer to demand can reduce distribution losses. Shortening the travel distance of electricity lessens the number of voltage transformation levels, and the freed capacity reduces utilization levels.

Options to Reduce Commercial Losses

The CL commonly associated with countries in the Latin America and the Caribbean region may occur for a variety of social, economic, and cultural reasons. Such losses—illicit uses, metering errors, and billing or administrative errors—may occur in the context of weak legal and institutional frameworks with poor enforcement of laws. In addition, utilities may lack the skills and technical resources needed to identify and control losses, adequate metering and information systems, and incentives to detect and combat electricity theft and audit the inventories of unmetered energy.

In areas with high concentrations of CL, such as zones with specific socioeconomic problems, the challenge for the utility is to implement a well-designed, comprehensive services and inspection plan that includes installation of connections and metering. To be effective, such efforts must be supported by government agencies working in tandem, often providing complementary social services and law enforcement or simply maintaining order.

In areas with low CL, it is especially difficult for the utility to detect irregular service, identify the causes of isolated cases, and take action to correct them on a case-by-case basis. Eventually, the cost of reducing losses may exceed the benefit, at which point it becomes preferable to tolerate a low CL level.

Demand-Side Efficiency

The counterpart to improving supply-side efficiencies in power production and distribution is a demand-side strategy that reduces peak and non-peak demand of end users. This, in turn, reduces the generation capacity and transmission and distribution assets needed to supply the system. Measures to reduce peak demand tend to be more popular with utilities than energy-efficiency measures per se since the former reduce their costs while the latter also reduce their income.

End-Use Energy Overview

Estimating the potential for energy efficiency begins by identifying how energy is currently used by the sector consuming it and the type of intended end use (e.g., lighting or air conditioning). The statistics on broad sector use, summarized in figure 7.3, show large differences between country shares across sectors owing to such factors as varying resource endowment, level of development, and degree of industrialization.

Existing information on energy end uses and subsectors for the two subregions is spotty; thus, it is often difficult to compare country data. The lack of systematically organized information is even greater for the Caribbean than Central America.

Potential for Central America and the Caribbean

The assessment of energy-efficiency potential is not trivial, especially in countries where such programs are just getting under way. This is the case for most countries in the subregions; thus, only a few preliminary estimates of national-level potential have been made.

For Costa Rica, it was estimated that the cumulative reduction in baseline electricity consumption over the 2002–16 period could reach 16 percent of total consumption. Since the impact of energy-efficiency policies grows over time, this implies that the reduction in annual consumption in 2016 would exceed 16 percent (CONACE 2003). For Panama, a study prepared in 2004 and later updated, taking 2009 as the base year, estimated that consumption could be reduced by about 10 percent by 2019 and 16 percent by 2023 (ECLAC 2009).

Figure 7.3 Sector Share of Total Electricity Consumption (GWh) for Selected Countries, 2007

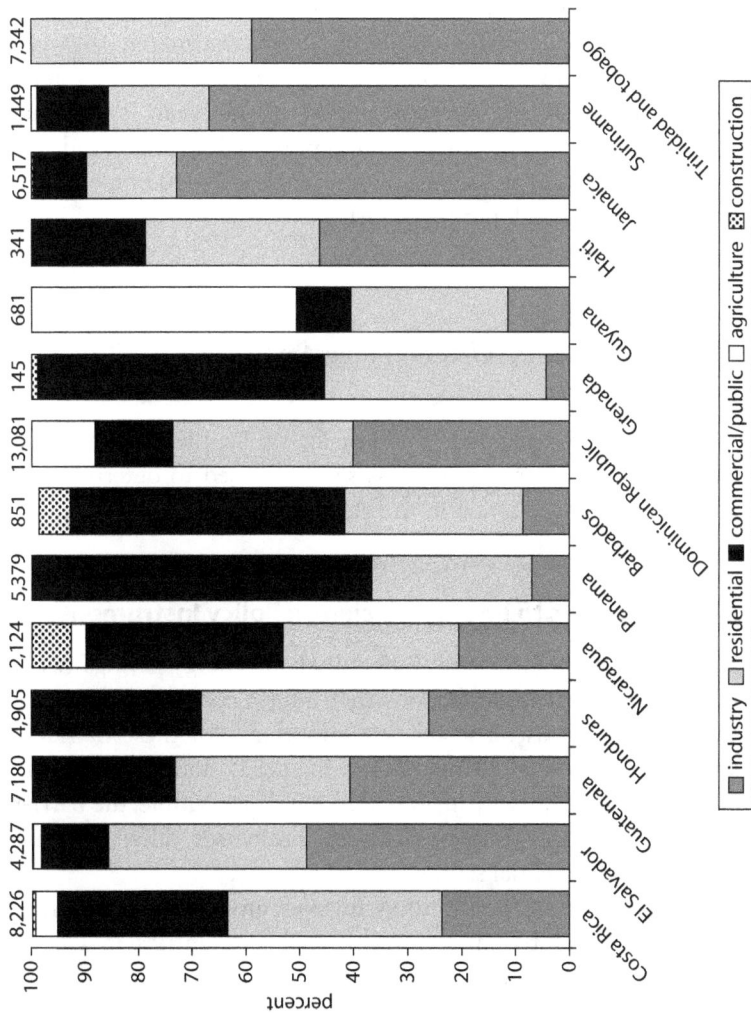

Legend: ■ industry ▨ residential ■ commercial/public □ agriculture ⊠ construction

Source: OLADE 2008.
Note: Values are calculated from the national energy balances, which are shown in barrels of oil equivalent (boe); 2006 values are reported for El Salvador and all Caribbean countries, except the Dominican Republic and Trinidad and Tobago.

Evaluations of specific market segments also point to significant potential savings. For example, audits conducted in three water-supply systems in El Salvador found a potential reduction in energy requirements of 30–40 percent. In Nicaragua, the potential reduction in electricity consumption and costs for some government departments was evaluated, focusing on fairly straightforward measures to improve lighting efficiency and air conditioning; it was estimated that efficiency measures would reduce total electricity consumption by about 27 percent, with a simple payback period of 2.4 years on average. Various audits conducted in Panama found potential savings in industry, commerce, and services (including government) of 27–40 percent, with an average simple payback period of just over two years for equipment investments.

Better Energy Efficiency: Effect on Investment and Oil Consumption

The method used to calculate the impact of supply-side efficiency improvements on oil consumption was similarly used to determine the effect of a 10 percent reduction in the electricity consumed (figure 7.4).

Overcoming Barriers to Energy Efficiency: Policy Instruments

Having a large economically-viable potential for energy savings results from numerous market imperfections that inhibit consumers' ability to optimize their energy use. The relative importance of these barriers— informational, institutional, technical, and financial—and their influence on consumer decisions vary by market segment. For example, the barriers to energy-efficiency investments faced by businesses differ markedly from those of the residential sector.

The most appropriate policy tools to overcome barriers to energy efficiency often differ by market segment. For countries in Central America and the Caribbean, the basic policy recommendations are as follows:

- *Deepen surveys and analyses of energy use and the factors influencing it.* These should be done for stocks of energy-using equipment, saturation rates for appliances, floor area of commercial buildings, and degree of days of cooling; having such an information base is important not only

Figure 7.4   Potential Fuel Savings in Electricity Generation from 10 Percent Improvement in End-Use Efficiency

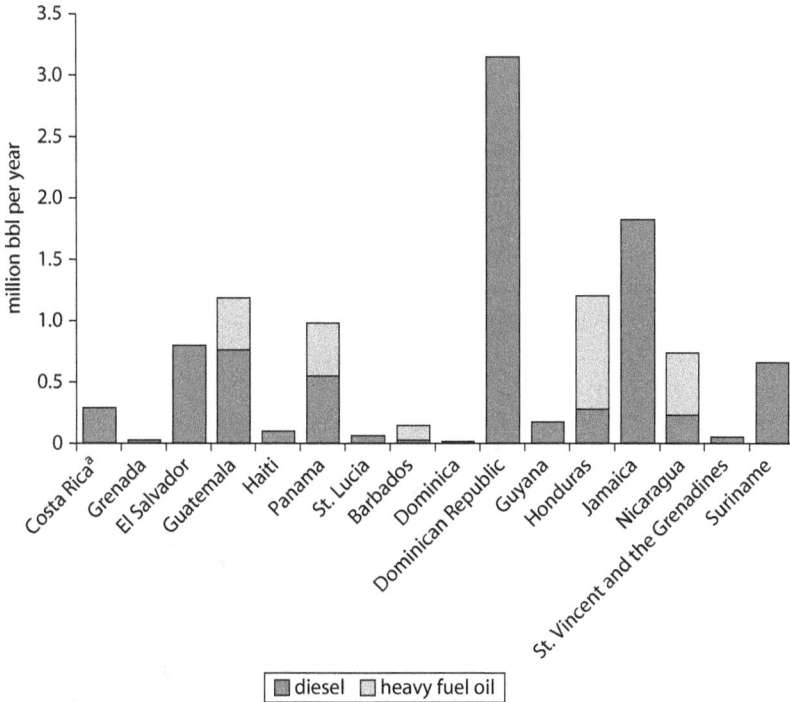

Source: Aburto 2010.

a. Costa Rica's small fuel consumption for electricity generation means that ascribing all GWh savings to reduced oil consumption exceeds total oil use for power generation. It has been somewhat arbitrarily assumed that the reduction in fuel consumption would be the same as for the supply side (figure 7.2). Combining the two values would imply about 60 percent reduction in fuel consumption, which may be close to the national system's operational limits; this suggests, for example, that a portion of diesel could be used in isolated off-grid systems, which would continue to operate.

for designing and implementing programs, but also for making credible estimates of their impacts on energy use.

• _Conceptualize energy-efficiency programs and policies as steps in a long-term process._ Policies and programs have short-term effects, but the impacts are cumulative, involving changes in both equipment and the behavior of many consumers.

• _Consider the institutional capabilities of the range of diverse stakeholders involved._ These typically include government agencies, project developers, and financial institutions.

- *Use labeling and standards programs for electric appliances intended for residences and small businesses.* These can be quite effective and have relatively low program costs.

- *Carefully design and implement appliance substitution programs.* Many countries are considering such a short-term measure for replacing outdated and inefficient equipment; such a program must be designed to encourage the participation of households, especially low-income ones, and small businesses without engendering onerous subsidy costs. Of course, old appliances must be properly disposed of.

- *Organize CFL replacement programs.* Compact fluorescent lamp (CFL) programs tend to be one of the first large-scale, energy-efficiency measures taken; they are relatively simple to organize and have been effective in many countries throughout the world. Currently, they are being tried in many countries of Central America and the Caribbean, and there is scope for program expansion in the region.

- *Familiarize banks with energy-efficiency lending to reduce perceived risk.* Lack of access to commercial financing has been a major impediment to expanding the market for energy-efficiency retrofitting projects. Banks are not accustomed to this type of project lending; they do not accept receivables from performance contracts as collateral, and are uncomfortable lending to project developers, such as energy services companies (ESCOs), which are usually poorly capitalized. A step-by-step process is needed to familiarize banks with this market to reduce perceived risk, which can enable the adaptation of loan-evaluation criteria and possibly the design of appropriate instruments. The World Bank and other donors can play a key role in this process by bringing international experience to bear and helping to establish and fund transitional mechanisms.

- *Carefully evaluate strategies that rely on permanent subsidies.* The price of energy is a crucial signal to consumers. Price distortion, the most common form of which is subsidy, makes energy rationalization more difficult. In most Central American countries and some Caribbean ones, governments have sought to avoid a full pass-through of the increased cost of energy supply to consumers, especially residential ones. The cost of these subsidies to the national treasuries, or the utilities forced to absorb them, has been quite high. To be effective, policies should have as their final objective the transformation of the targeted market segment, such that a new level of energy-efficiency performance is reached on a sustainable basis. In general, governments should assign a higher priority to poverty reduction policies over energy-product subsidies to ensure that prices reflect the costs of providing the goods and services or their international benchmarks. In cases where the government finds it necessary to support low-income consumers, it would be preferable to introduce income-enhancing measures that directly target poor households.

Summary Remarks

This chapter has demonstrated the potential fuel savings that can result from energy supply-side and end-use efficiency enhancements. Greater supply-side efficiency can be achieved by reducing technical losses, which depends on modifying system characteristics and configurations. The marginal effect of commercial losses can also be reduced through social-services and law-enforcement measures. Higher demand-side efficiency can be achieved through labeling, norms and minimum standards for appliances and buildings, and information programs and demonstrations. Financial deepening, in the form of credit programs, is also required to ensure companies' effective involvement in end-use efficiency. Among the countries studied, Jamaica and the Dominican Republic exhibit the largest estimated fuel savings. The next chapter considers the potential of a third structural measure—regional energy integration—to mitigate medium- and longer-term vulnerability to high and volatile oil prices.

Notes

1. Studies on the economics of climate-change mitigation show that energy-efficiency measures are the most cost-effective way to reduce greenhouse gas (GHG) energy emissions. In fact, since their benefits usually outweigh their costs (without factoring in transaction costs), they are usually reported as having a negative cost per ton of CO_2 (Johnson et al. 2009).

2. Statistical data were collected from the United Nations (UN), U.S. Energy Information Administration (EIA), Latin American Energy Organization (OLADE), UN Economic Commission for Latin America and the Caribbean (ECLAC), Caribbean Electric Utility Service Corporation (CARILEC), Regional Energy Integration Commission (CIER), Central American Electrification Council (CEAC), and country sources. For the purpose of this

analysis, the three larger databases—those of the UN, EIA, and OLADE—
were examined and compared.

3. Efficient losses are based on data from Puerto Rico (for small island countries,
with total losses of 6.9 percent), and Chile (for larger countries, with total
losses of 7.9 percent). For countries with high total losses, the reference TL
were fixed at 9.4 percent (for larger countries) and 8.4 percent (expert
assumption for small island countries). Reference CL were defined as the difference between total and reference TL. However, for countries with total
losses less than or slightly exceeding the reference TL, the reference CL were
adjusted to 2 percent or slightly less. Thus, for these countries, the reference
TL became the remainder. Fixing the reference CL at 2 percent or slightly
lower resulted from the perception that CL are usually present in LAC countries for cultural, social, and economic reasons. Also, it is difficult to identify
isolated CL cases, and further reducing them is unlikely to be cost-efficient.

References

Aburto, J. 2010. *Study on Mitigating Impact of Oil Price Volatility in Central
America and the Caribbean Energy Efficiency of Public Utilities: Transmission
and Distribution Losses.* Report prepared for the World Bank, Washington,
DC: World Bank.

CONACE (National Commission on Energy Conservation). 2003. *Programa
Nacional de Conservación de Energía 2003–2008 (PRONACE).* San José:
Comisión Nacional de Conservación de Energía.

ECLAC. 2009. "Situación y Perspectivas de la Eficiencia Energética en América
Latina y El Caribe." Prepared for the Regional Intergoverment Meeting on
"Energy Efficiency in Latin America and the Caribbean," Santiago, Chile.
September 15–16.

Johnson, T. M., C. Alatorre, Z. Romo, and F. Liu. 2009. *Low-carbon Development
for Mexico.* Washington, DC: World Bank.

OLADE (Latin American Energy Organization). 2008. *Energy Statistics
Report—2007.* Quito: Organización Latinoamericana de Energía.

CHAPTER 8

Regional Energy Integration

Regional energy integration can help countries to diversify their power-generation mix. Having more diversified generation sources, in turn, can reduce variable costs; as more renewable energy sources and natural gas enter the generation mix, greenhouse gas (GHG) emissions can be reduced. Regional integration with countries that have more diversified generation matrices can lower the total share of imported oil and oil products in a country's generation mix and thus mitigate its vulnerability to high and volatile oil prices.

This chapter addresses some of the regional integration opportunities that can help countries in Central America and the Caribbean to reduce their vulnerability to high and volatile oil prices. The next section focuses on these subregions' potential for electricity interconnection. This is followed by a discussion of their potential for natural gas integration.

Regional Integration in Electricity

Country and regional interconnections allow for the optimization of electricity supplies, which can improve efficiency and reduce expenses in high-cost generation capacity. Owing to economies of scale, integration leads to lower generation costs. And when the consumption profiles of participants are not perfectly correlated, the smoother load pattern that arises means less investment in reserve requirements. If these conditions are met, fossil fuel use decreases along with the countries' vulnerability to high and volatile oil prices. Furthermore, from a market perspective, regional integration promotes competition and helps to realize trade gains associated with specialization of the most efficient producers.

While the economic benefits of an integrated market are generally accepted, a series of institutional obstacles often prevents the formation of regional exchanges. The most common problems are use of multiple technology standards; varying regulatory regimes, legal frameworks, and pricing policies; and environmental concerns. Additional hurdles that can limit or delay market integration include perspectives at variance with shared investment costs, uncertainties in political decision-making, and lack of a critical mass to enable investments in non-oil-fired power generation. The subsections below highlight the challenges for Central America and the Caribbean, some of which are unique to these subregions' geography and energy systems, as well as the scope and types of data and analyses available.

Central America

The electricity interconnection system in Central America was initiated in the early 1990s. Interest had been spurred mainly by the desire to (i) enhance the security of small-sized national networks and (ii) take advantage of the relatively large share of hydroelectric power in the supply mix. Larger hydropower projects could more easily be absorbed into a regional, versus a national, market since linking diverse hydrological basins could help to lessen the impacts of annual and seasonal output variations; the substantial potential for developing new hydro-generation capacity was also recognized.

Efforts to promote integration in Central America resulted in the Central American Electrical Interconnection System, more commonly known as the SIEPAC Project. SIEPAC interconnects all six countries in the subregion (Costa Rica, El Salvador, Guatemala, Honduras, Nicaragua, and Panama) (map 8.1). From the outset, the SIEPAC transmission line was designed to create a regional electricity market (MER), for which a framework treaty was signed by the six countries in 1996; two years later, the treaty came into force. MER regulations were first applied in 2002 to the existing Honduras-El Salvador interconnection.

Construction under the SIEPAC Project began in 2006. It has included approximately 1,830 km of 230-kV transmission line extending from

Map 8.1   SIEPAC Regional Transmission Line

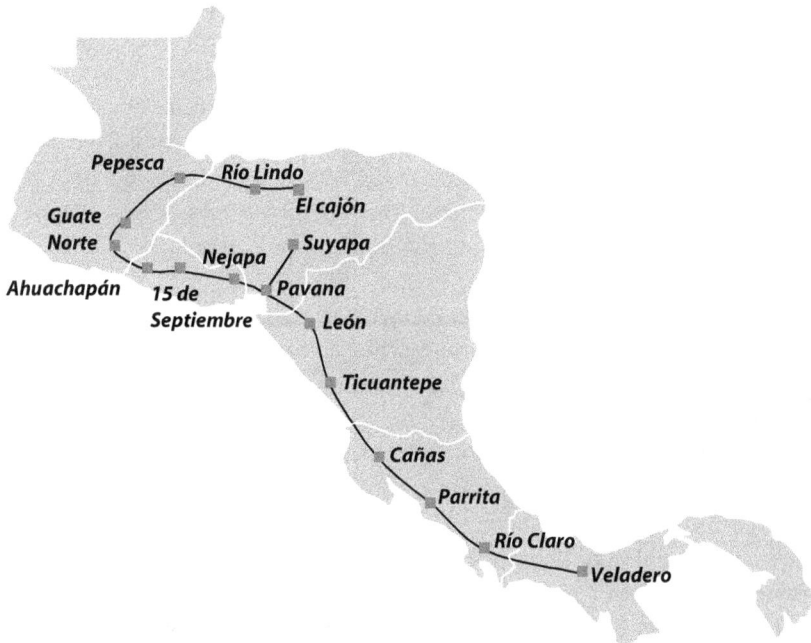

Source: SIEPAC data.

Guatemala to Panama, as well as connections and transformation substations in each of the six countries. The MER is expected to be fully operational by late 2012, and the program is now considering expanding capacity for steady interchanges of up to 300 MW.

SIEPAC aims to promote greater energy trade in Central America by (i) reducing the region's relatively high electricity costs, (ii) enabling implementation of large-scale hydropower and natural-gas combined-cycle projects, and (iii) strengthening the reliability of national electricity systems. As a result of the project, the six Central American countries will enjoy a better investment environment that facilitates the financing of larger projects. Once installed, the interconnection will enable new, larger projects to take advantage of economies of scale in electricity generation. According to the Central American Electrification Council (CEAC), within 8–10 years, SIEPAC could realize operational cost savings of about 4 percent and fuel savings of about 3 percent, based on indicative expansion planning exercises (World Bank 2011).

One SIEPAC objective is to create a commercial and regulatory structure that allows the six countries to gradually progress toward harmonized internal regulations; such institutional arrangements are permitted to occur before completion of the new infrastructure. Progress on harmonization has been slowed by deep institutional differences, and a chronic shortage of generating capacity within countries has led to a decline in intra-regional trade (table 8.1). At the same time, mechanisms that coordinate electricity purchases through firm energy contracts may help to spur greater investment in renewable power generation and diversify the energy matrix across the region.

Without increasing generation capacity, there is concern that the new SIEPAC transmission infrastructure will be under-utilized. Results of a quantitative exercise undertaken for Central America show that, by relying on hydroelectric plants that could be built in the subregion, greater integration would increase hydro share by 8 percent (from 46 to 54 percent), resulting in a 14 percent reduction in CO_2 emissions from reduced use of thermal power (figure 8.1). By lessening the need for reserve capacity, significant savings could be realized from integration in domestic power-sector investment (Yépez-García, Johnson, and Andrés 2011).

This exercise shows how a larger share of hydropower generation, owing to greater electricity integration, creates a more diversified energy matrix in Central America and Panama. More regional integration makes larger projects with considerable scale economics possible. As such projects enter the regional generation matrix, countries lessen their dependence on oil imports and reduce their vulnerability. The simulation in this

Table 8.1   Evolution of Intra-Regional Electricity Trade
Imports plus exports (GWh)

Year	Costa Rica	El Salvador	Guatemala	Honduras	Nicaragua	Panama	Central America
2000	497.7	919.4	963.8	300.6	117.0	147.4	2,945.9
2001	240.7	396.6	422.0	308.6	17.3	160.9	1,546.1
2002	476.1	494.7	485.3	415.1	22.1	83.7	1,977.0
2003	160.1	530.3	446.9	336.8	33.1	183.7	1,690.8
2004	394.0	549.6	505.1	392.2	45.1	285.1	2,171.2
2005	151.0	359.9	358.6	61.1	30.8	161.2	1,122.8
2006	130.0	21.5	96.6	18.0	53.4	117.0	436.0

Source: ECLAC and SICA 2007.

Figure 8.1   Impact of Electricity Integration in Central America

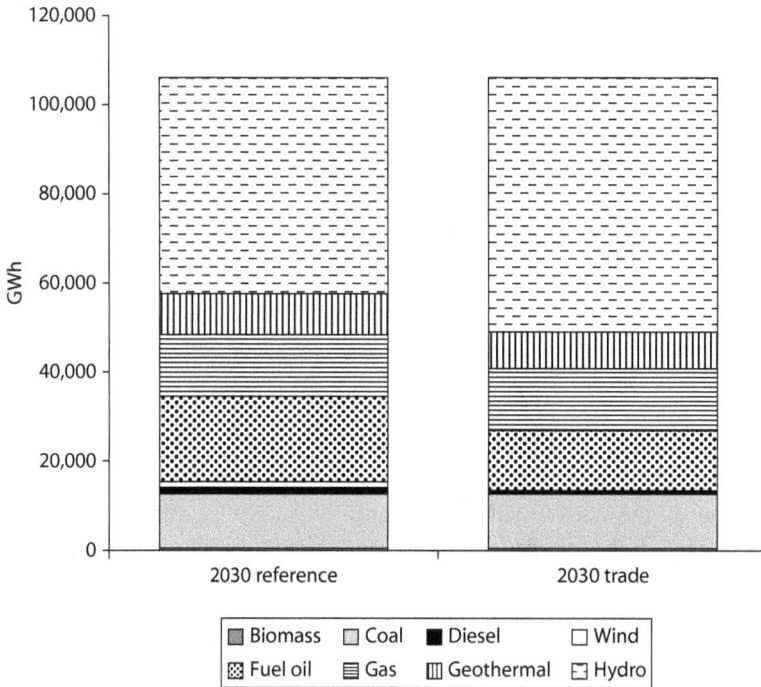

Source: Authors, based on optimization model.

exercise was limited to a certain amount of additional hydro potential in Central America and thus does not comprise hydro's full potential for diversifying Central America's generation matrix. Another option would be to tap more hydro markets in North and South America, particularly those in Mexico and Colombia.

Concurrent with SIEPAC implementation, a new framework, known as the Mesoamerican Energy Integration Program (PIEM), has been put in place to develop extra-regional integration initiatives.[1] Under PIEM, the plan is to connect the SIEPAC line with Mexico and Colombia. These countries would become MER members via the Mexico-Guatemala and Colombia-Panama interconnections, described below:

- *Mexico-Guatemala interconnection*: This 400-kV transmission line of 103 km (32 km in Mexico and 71 km in Guatemala) recently became operational. The initial transfer capacity is 200 MW from Mexico to Guatemala and 70 MW from Guatemala to Mexico. Transactions can be made with other Central American countries through the SIEPAC line.

- *Colombia-Panama interconnection*: Now in the advanced planning stages, this interconnection is to have 514 km of high-voltage direct current (HVDC) transmission line (250 or 450 kV), including a submarine link between the Cerromatoso (Colombia) and Panama II substations, requiring an investment of about US$200 million (WEC 2008). Initial capacity would be 300 MW from Colombia to Panama (with a possible 600-MW expansion capacity) and 200 MW from Panama to Colombia.

Both Mexico and Colombia are expected to have spare capacity in the near future to export power to Central America; in a relatively short time, these two interconnections, particularly the Colombia-Panama one, could significantly impact the supply-demand balance of countries in the sub-region. The Colombia-Panama interconnection could help to consolidate use of the new SIEPAC infrastructure, which would improve the potential for effectively developing some of the region's larger hydro, as well as geothermal and wind, potential. Looking ahead, Central America might become a corridor for a more robust interconnection between Colombia and Mexico. The extra-regional interconnections should bring significant short-, medium-, and long-term benefits and open new avenues for regional exchanges.

Caribbean Islands

The potential for electricity interconnection in Caribbean nations is more limited than in Central America, given that these islands are located far from each other and have smaller markets. Even so, electricity interconnection could significantly reduce these countries' dependence on oil-fired generation. Various cases show that interconnections between two or more countries could be economically feasible, taking advantage of economies of scale and development of indigenous resources.

Some islands of the Lesser Antilles have substantial conventional geothermal potential. The most significant resources are in Nevis and Dominica, while Guadeloupe and Martinique also have possibilities (Guadeloupe has a 4-MW pilot plant). The potential for geothermal exploitation depends on two sets of possible interconnections. The first, which is highly economic, is located between Nevis and St. Kitts (Nexant

2010); while the second links Gaudeloupe and Martinique to the superior geothermal resources in Dominica. Another possibility is an interconnection between the Dominican Republic and Haiti.

The interconnection of electricity markets in the subregion could be achieved using either submarine cables between islands or land transmission lines between the Dominican Republic and Haiti (box 8.1). Electricity interconnections make sense between island nations, particularly when the countries take advantage of economies of scale, such as for natural gas or geothermal technologies. Various options have been

---

Box 8.1

Promoting Energy Integration in the Caribbean

A recent study by Gerner and Hansen evaluated potential opportunities for energy integration in the Caribbean, considering both the technical and economic viability of the options considered. The authors identified three major areas for development, as follows:

Renewable energy. Resources found to have the greatest interconnection potential are natural gas (pipeline and liquefied natural gas [LNG]), geothermal, wind, small hydropower, and biomass. All are highly competitive with technologies currently in use. A key challenge is to identify sites with good resources that are economically feasible.

Electricity interconnections using submarine cables. Interconnecting the various islands using submarine cables would improve efficiency and increase electricity-sector security. Also, it would enable more large-scale energy generation using renewables. The level of interconnection could be subregional, continental (e.g., with Mexico, Colombia, or República Bolivariana de Venezuela), or bilateral (e.g., Montserrat-Antigua and Barbuda or Puerto Rico-the Dominican Republic).

Gas pipeline interconnections. The study finds that supplying natural gas through the proposed Eastern Caribbean Gas Pipeline might be cheaper than current diesel-based generation. Natural gas from Trinidad and Tobago would supply Barbados, Guadeloupe, Martinque, and St. Lucia. If the islands are interconnected, the pipeline could take advantage of economies of scale owing to the large volumes of gas transported. To be implemented, however, the project must first win consensus among diverse stakeholders, ranging from gas suppliers, utilities, and regulators to financial institutions and governments.

Sources: Gerner 2010; Gerner and Hansen 2011.

evaluated as technically feasible and economically viable (Gerner and Hansen 2011). Such interconnections as Dominica-Martinique, Dominica-Guadaloupe, Nevis-St. Kitts, Saba-St. Maarten, and the Dominican Republic-Haiti offer significant economic benefits and thus have enormous potential for development.

Natural Gas

The rapidly changing world market for natural gas favors the fuel's entry in the supply mix of many countries in Central America and the Caribbean. As discussed in chapter 6, use of natural gas would help countries to further diversify their electricity generation mix. Moreover, because of the recent decoupling of natural gas prices from those of oil, natural gas could play a key role in mitigating vulnerability to higher oil prices over the medium term.

Expanded use of natural gas would require developing pipelines for intra-regional integration and building LNG re-gasification plants to permit countries to integrate into international markets. The resource's scant availability in most countries of Central America and the Caribbean means that most of it would have to be imported from outside the two subregions.

Central America

Guatemala is the only Central American country that produces associated gas, albeit in small quantities; the other five countries lack both associated and non-associated gas reserves. The concept of introducing natural gas to the subregion as a means of diversifying the energy matrix was first considered in 1996. An early proposed option was to build a 2,300-km isthmus gas pipeline extending from Mexico to Panama, the construction of which would be part of the regional agenda for energy integration. An initial exercise to identify the project's economic benefits found that having to import natural gas from a net importer, such as Mexico, would limit its viability; however, this issue opened discussion on evaluating alternate options, including the construction of an LNG plant. However, both an isthmus pipeline and an LNG plant would require large investments and unprecedented synchronization of national expansion plans in order to build a series of power plants using natural gas to meet demand needs.

El Salvador has explored the possibility of an LNG re-gasification terminal; to date, however, the project has been delayed for lack of a critical mass of demand and financial resources. In 2005, Panama and Colombia agreed on a new proposal, whereby Colombia would export natural gas to Panama. Both pipeline and LNG options are being considered (WEC 2008). But the Colombia-Panama proposal competes directly with the electricity interconnection, whose planning is further advanced.

A more viable option might be to promote the entry of natural-gas power plants in two or more countries to achieve minimum economies of scale for an initial LNG re-gasification terminal. This would require a fraction of the effort needed for the isthmus pipeline and would lay the groundwork for new pipeline growth. The northern-tier countries of Guatemala and El Salvador, together with Honduras, may present the most promising target for an LNG terminal, whether along the Pacific or Atlantic coast. The impact of the Panama-Colombia-electricity interconnection would be less on these countries than on southern-tier ones. The gains from such a project would be relegated oil-fired capacity and a lower capacity factor reserve because of a smoother combined load demand.

Caribbean Islands

For the Caribbean island nations, the expanded use of natural gas could contribute substantially to reducing dependence on oil derivatives in the short and medium term. Significant reserves in the subregion are found only in Trinidad and Tobago, where natural gas has long been the fuel source for nearly all electricity generation. The challenge has been how to move gas to the other islands, which are located far from each other and whose market size is small. Of the two available approaches, pipelines and an LNG terminal, only the latter has been tried to date, first in Puerto Rico in 2000 and in the Dominican Republic two years later. In both cases, the LNG terminal was developed jointly with a power plant (540 MW in Puerto Rico and 320 MW in the Dominican Republic). Both projects have succeeded both technically and economically.

Today, the Caribbean has no inter-island gas pipelines. However, a pipeline concept was proposed by the prime minister of Trinidad and Tobago in 2002, and a partnership, called the Eastern Caribbean Gas Pipeline Company (ECGPC), has been formed. The proposal is to construct a pipeline from Tobago to Barbados and then on to Martinique and Guadeloupe. The initial section (from Tobago to northwest of Barbados) would consist of a 172-mile, 12-inch pipeline with a small offshore lateral to the main power plant. The second section (from Barbados to Martinique) would be a 120-mile, 10-inch line with a side spur to St. Lucia. The third section (from Martinique to Guadeloupe) would be a 188-mile, 8-inch line. The pipeline is designed to send 50 MMSCFD to Barbados and 100 MMSCFD combined gas to Martinique, Guadeloupe, and St. Lucia.

Clearly, as mentioned above, there is potential for integration between the Dominican Republic and Haiti. Though historically relations between the two countries have been strained, Haiti's urgent need for reconstruction following the 2010 earthquake may offer an opportunity to overcome differences, with substantial benefits to both sides. The Dominican Republic, which has the most diversified energy generation mix in the Caribbean, has had a positive experience with its LNG terminal and natural-gas power plants; yet it still suffers from high costs and an unreliable electricity grid. Haiti's power system was quite underdeveloped even before 2010, with a per-capita grid supply less than one twenty-fifth that of the Dominican Republic.

Thus, the short-term development needs of the electric power system extend far beyond simple reconstruction. An obvious approach would involve two forms of energy interconnection and trade between the two countries. First, a pipeline would need to be constructed from the existing LNG terminal in the Dominican Republic to Haiti. This would make it possible to build a natural-gas power plant, which would cost far less than the alternative of small distillate power plants. Second, the pipeline might be complemented by an electricity interconnection between the two countries to increase reliability for both.[2]

Conclusion

This chapter has outlined the various approaches tried in the ongoing effort to develop an integrated power market in Central America and the Caribbean. The two main mechanisms are constructing electricity interconnections and building natural-gas infrastructure (in the form of new LNG terminals or pipelines and new plants). As an integrated power market, Central America leads in experience and level of progress. Its advanced integration plans to trade electricity with Mexico in the north and Colombia in the south offer a clear path to reducing the subregion's vulnerability to higher and more volatile oil prices. In the Caribbean, the geothermal potential of some island nations can serve as the basis for a more diversified power market that is less vulnerable to oil prices. The Dominican Republic and Haiti, in particular, can benefit from stronger integration on both the power and natural-gas fronts.

Clearly, developing an integrated electricity market is an important medium- and long-term strategy for reducing vulnerability to high and volatile oil prices in Central America and the Caribbean. Integrated markets allow for fuel savings through a more diversified power mix (in the case of these subregions, hydroelectric and thermal plants become economically viable), economies of scale, and reduction for reserve capacity. Moreover, all such benefits imply a reduction in GHG emissions. The persistent constraints in these subregions mainly involve heterogeneous regulatory regimes and a politically uncertain environment for enabling projects over the long term.

Notes

1. In December 2005, PIEM was adopted under the Declaration of Cancun by the governments of Belize, Colombia, Costa Rica, the Dominican Republic, El Salvador, Guatemala, Honduras, Mexico, Nicaragua, and Panama.

2. One recent analysis of the Dominican Republic-Haiti electricity interconnection was unfavorable (Nexant 2010); however, it assumed that electricity exported from the Dominican Republic to Haiti would be based only on fuel-oil generation.

References

ECLAC (UN Economic Commission for Latin America and the Caribbean) and SICA (Central American Integration System). 2007. "Estrategia Energética Sustentable Centroamericana 2020." UN Economic Commission for Latin America and the Caribbean and General Secretariat of the Central American Integration System, Mexico City.

Gerner, F. 2010. *Caribbean Regional Electricity Generation, Interconnection and Fuels Supply Strategy: Synthesis Report.* Prepared with support of the LCR Energy Group. Washington, DC: World Bank.

Gerner, F., and M. Hansen. 2011. *Caribbean Regional Electricity Supply Options: Toward Greater Security, Renewables, and Resilience.* Report No. 59459. Washington, DC: World Bank.

Nexant. 2010. *Caribbean Regional Electricity Generation, Interconnection and Fuels Supply Strategy: Interim Report.* Prepared for the World Bank (January). San Francisco, CA: Nexant.

WEC (World Energy Council). 2008. *Regional Energy Integration in Latin America and the Caribbean.* London: World Energy Council.

World Bank. 2011. *Regional Power Integration: Structural and Regulatory Challenges.* Central America Regional Programmatic Study for the Energy Sector, Report No. 58934-LAC. Washington, DC: World Bank.

Yépez-García, R. A., T. M. Johnson, and L. A. Andrés. 2011. *Meeting the Balance of Electricity Supply and Demand in Latin America and the Caribbean.* Directions in Development. Washington, DC: World Bank.

CHAPTER 9

How Much Can It Help?

The strategies presented in the previous chapters are highly complementary. As evidenced in chapter 5, an array of financial instruments might be used to protect against short- and possibly medium-term oil price volatility. But in the long run, mitigating vulnerability to high and volatile oil prices depends on structural measures designed to reduce a country's oil consumption. As a synthesis exercise, this chapter attempts to quantify the potential gains that can accrue to the power sector from implementing the three structural measures presented in chapters 6, 7, and 8—more intensive use of renewable energy sources, improved energy efficiency, and greater regional integration with countries less vulnerable to oil prices. While the calculations refer to Central America and the Caribbean, the underlying principles of the policy recommendations can be applied to any oil-importing country seeking to mitigate oil price risk.

Review: Effects of High and Volatile Oil Prices

As discussed in chapter 2, high and volatile oil prices have far-reaching effects, both direct and indirect, extending to firms, households, government, and the overall competitiveness and financial viability of the national economy. To reiterate, the major direct effects are (i) a deteriorating trade balance, through a higher import bill, reflecting a worsening in terms of trade; (ii) a weakening fiscal balance, due to greater government transfers and subsidies to insulate movements in international energy markets; and (iii) investment uncertainty, resulting from the higher risk of engaging in new projects and associated development and sunk costs.

The major indirect effects are (i) headline inflation, which may feed into core inflation through rising inflation expectations that trigger wage spirals; (ii) loss of consumer confidence and purchasing power, due to greater economic uncertainty and higher inflation, which may reduce household discretionary spending and thus affect a major component of the economy; (iii) loss of competitiveness from higher power generation and transport costs, leading to decreased international competitiveness; and (iv) institutional weakening, as firms and households pressure the government to bypass market mechanisms, which, in turn, affects the credibility and functioning of the regulatory environment.

Reducing Oil Dependence

The benefits of implementing more intensive use of renewable energy in the generation mix, higher energy efficiency in both supply and demand, and greater regional integration with more energy-diversified countries can be estimated as a result of the reduction in oil consumption. The associated savings are summarized in figures 9.1a and 9.1b.

More Renewables in the Generation Mix

As discussed in chapter 6, renewable fuels directly reduce the need for oil-based fuel as a source of power generation. Such substitution also reduces greenhouse gas (GHG) emissions. Figures 9.1a and 9.1b underscore the role of renewables in reducing the countries' exposure to higher and volatile oil prices. Among the long-term structural strategies, a 10 percent increase in renewable potential capacity results in larger fuel savings for most countries. For both Central America and the Caribbean, greater use of renewables could lead to savings of 14.2 million and 5.6 million barrels of diesel and heavy fuel oil (HFO), respectively. These savings represent an average reduction of 1.66 percent of GDP.

Greater Energy Efficiency

Greater energy efficiency is reflected in higher power output for a given amount of fuel or from lower fuel requirements to generate a fixed

Figure 9.1   Summary of Potential Fuel Savings from Implementing Structural Measures

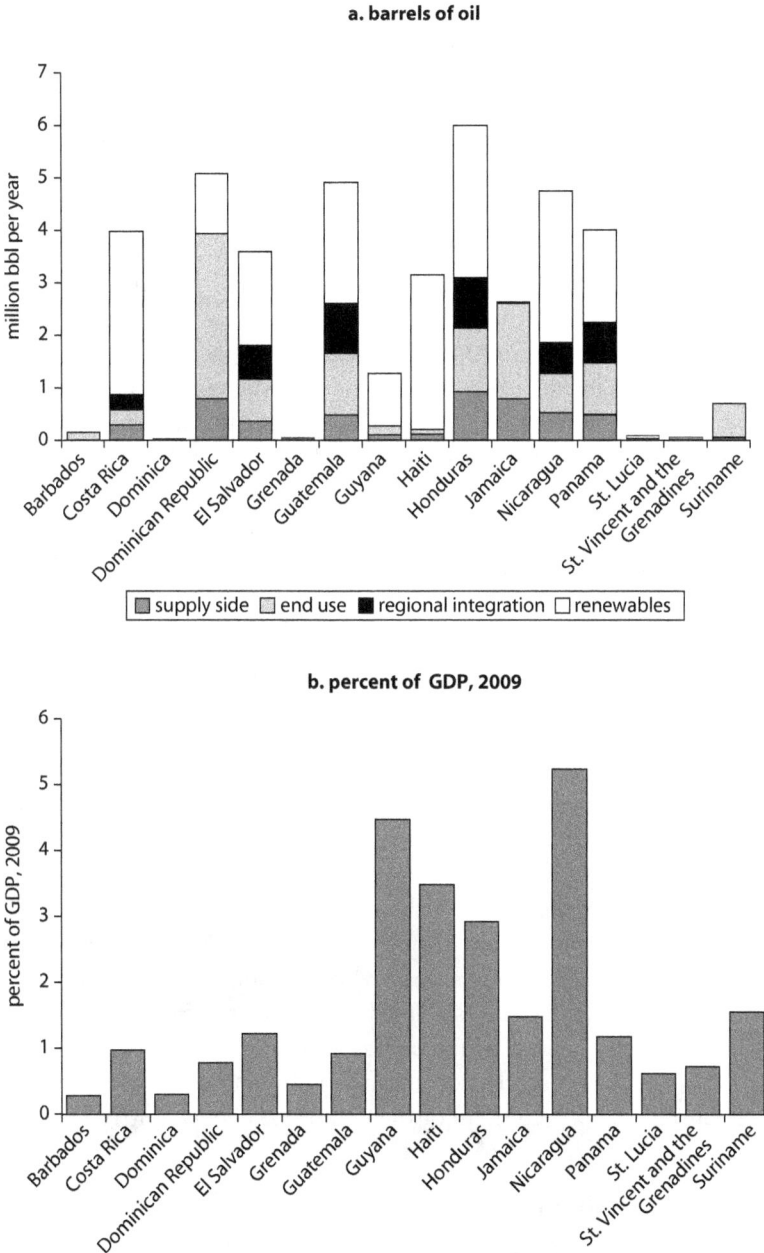

a. barrels of oil

b. percent of GDP, 2009

Source: Authors; based on data from Aburto 2010, OLADE 2008, and Nexant 2010.

amount of electricity. The associated savings occur on both the supply (production, transmission, and distribution) and demand (end use) sides. The first stage of potential efficiency gains derives from higher marginal efficiency in generation; that is, the amount of electricity obtained per unit of fuel in oil-fired generation plants. This often requires investing in newer, more efficient technologies or achieving larger economies of scale. Other technical losses (TL) are associated with transmission and distribution networks, where grid configuration, design, and obsolescence impact the natural losses from energy flows through the system.

Higher end-use efficiency entails (i) increasing consumption efficiency (related to the amount of electricity needed to power electric equipment and reduce demand) and (ii) reducing commercial losses (CL) (concerned with preventing illicit use, metering errors, and billing or administrative errors). While CL can be considered transfers from the utility to consumers and not necessarily related to reducing electricity consumption, they nonetheless distort information related to capacity planning and systems operation.

For nearly every country studied in Central America and the Caribbean, end-use savings are larger than supply-side ones. For these respective subregions, end-use savings amount to 9 million and 2.4 million barrels of diesel and HFO, compared to 3.5 million and 1.5 million barrels on the supply side.

Regional Integration

Regional power-market integration can lead to a more diversified energy matrix since countries vary in natural-resource endowments. It can also lead to economies of scale that favor the use of hydropower and other renewable fuels. The energy savings resulting from regional integration accrue from (i) regional interconnections, which optimize electricity supplies; (ii) less investment in reserves requirements; (iii) smoother load patterns from a larger aggregation of consumers; and (iv) promotion of competition.[1]

For Central America, the estimated annual savings from regional integration in electricity are 2.4 million barrels of diesel fuel and 1.8 million barrels of HFO. These figures suggest a reduction of approximately 8 percent in the oil-fired share of the countries' energy matrix.

What Can Be Gained

While such savings can mitigate the direct and indirect effects of higher oil prices, the impact varies by country, depending on each one's natural resources and energy matrix. Overall, however, the negative effects diminish. Predictably, the greater gains accrue in countries whose energy prices are subject to more distortions and are tightly linked to fiscal balances.

At both the macroeconomic and microeconomic levels, less exposure to international energy prices would lessen inflationary effects and the erosion of competiveness and consumer confidence. Likewise, less exposure to higher volatility in the international markets would facilitate longer-term energy planning and investment.

More directly, the fuel savings associated with using the three structural measures in a combined strategy can be measured in terms of the impact on the countries' current account. Net oil-importing countries would experience a lower fuel bill, which, in turn, would improve their current account. The combined monetary savings for the two subregions is estimated at US$3.7 billion. In 2008, the most recent year for which data is available for all of the countries studied, 19 countries (i.e., all but Suriname) were running a current account deficit; thus, the potential fuel savings would directly improve the current account. A parallel effect from such an improvement would be the release of foreign exchange for other uses or as central bank reserves (figure 9.2).

Figure 9.2   Impact of Potential Fuel Savings on Current Account for Selected Countries

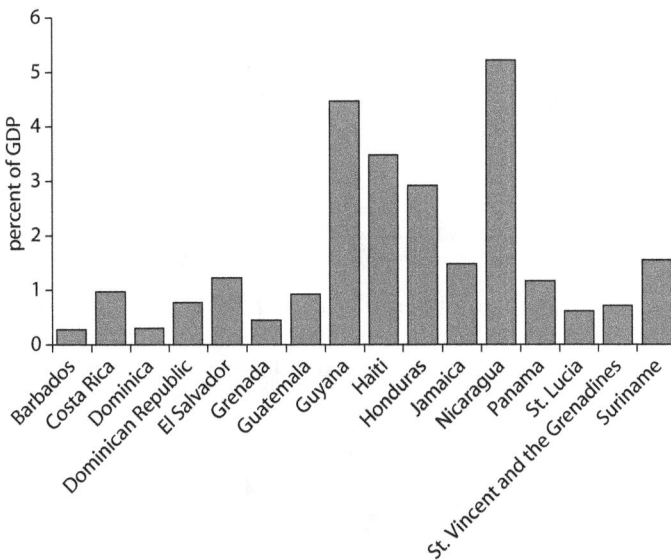

Assuming that the countries' current account relative to GDP remained unchanged, the average improvement in the current account balance would amount to approximately 1.66 percent of GDP. At the country level, Guyana and Nicaragua could witness a reduction of up to 5 percent of GDP in their current account deficit; while the reductions for Haiti and Honduras would be 3.5 and 2.9 percent of GDP, respectively.

Summing Up

Taken together, the structural measures to mitigate vulnerability to high and more volatile oil prices can result in significant savings in fuel purchases. The savings associated with more intensive use of renewables, higher-efficiency generation and use, and a more regionally integrated power market could amount to 29.1 million and 11.3 million barrels per year of diesel and HFO, respectively. These, in turn, represent the equivalent of US$2.9 billion, based on the 2009 average price for each of the fuels. Such savings would buffer net oil-importing countries from the adverse effects of upward trending oil prices and greater volatility in international energy markets. At the same time, it is important to keep in mind the considerable upfront costs to households, firms, and utilities that making such a structural transition may entail.

Note

1. In addition to the benefits of regional electricity integration, natural gas integration can help to diversify the generation matrix, along with decoupling prices from oil-derived fuels, which further reduces vulnerability to higher oil prices.

References

Aburto, J. 2010. *Study on Mitigating Impact of Oil Price Volatility in Central America and the Caribbean Energy Efficiency of Public Utilities: Transmission and Distribution Losses.* Report prepared for the World Bank, Washington, DC: World Bank.

Nexant. 2010. *Caribbean Regional Electricity Generation, Interconnection and Fuels Supply Strategy: Interim Report.* Prepared for the World Bank (January). San Francisco, CA: Nexant.

OLADE (Latin American Energy Organization). 2008. *Energy Statistics Report—2007.* Quito: Organización Latinoamericana de Energía.

CHAPTER 10

Conclusions

The combination of significant oil imports and consumption has made some countries extremely vulnerable to the oil price volatility observed in recent years. Oil-importing countries with a large share of oil in their energy mix are especially vulnerable to high and volatile oil prices. At both the macro and micro levels, their economies suffer numerous effects, the duration of which ranges from the short term to permanent changes that hinder potential growth and international competitiveness.

At the macro level, oil prices directly affect the aggregate economy. Directly or indirectly, they can have an immediate or lagged effect on government finances, as well as the balance of payments. As higher energy prices (of oil or power) are passed on to consumers, a series of responses may be triggered, including rising inflation expectations or, in the presence of energy subsidies, a deteriorating fiscal balance. At the micro level, high oil prices weaken the regulatory framework as governments implement nonmarket mechanisms to accommodate consumer demand for intervention. In addition, volatile oil prices affect the investment and consumption decisions of economic agents.

Results of the exercise presented in chapter 9 demonstrate that, by implementing the structural measures suggested in the preceding chapters (chapters 6–8), substantial savings can accrue to heavily oil-dependent countries. The components of this three-pronged strategy—a more diversified energy supply system, including greater utilization of available renewable sources; improved efficiency in electricity production and use; and regional integration, which promotes energy diversification—can work together over the long term to effectively reduce a country's oil generation and consumption and thus mitigate its vulnerability to high and volatile oil prices. At a macro level, less oil consumption can directly improve a country's aggregate economy and directly and indirectly benefit government finances and balance of payments. At a micro level, less vulnerability to oil price risk can facilitate investment planning and consumer decision-making.

Complementary to these structural measures, price risk management instruments may mitigate exposure to the shorter-term economic uncertainty created by oil price volatility, which also affects investment and planning decisions by households and firms. The risk of oil price volatility can be hedged by using financial instruments or incorporating price protection into physical contract mechanisms, such as long-term fixed forward pricing. As discussed in chapter 5, the use of commodity risk management tools is not widespread among governments, although such tools are well-established in the commercial sector. Before selecting a hedging strategy, it is critical for a country to undertake careful risk assessment and evaluate various hedging approaches against its unique needs.

Equally important is establishing an institutional framework that adequately supports implementation of a price risk management strategy. Key steps in the overall process of establishing a commodity hedging strategy are documenting the reasons for selecting a specific hedging product; establishing the roles and responsibilities of the various actors and agencies; adequately verifying the legal and regulatory infrastructure; establishing procedures for selecting counterparties and brokers; and exercising careful oversight, supervision, and reporting.

In the case of Latin America and the Caribbean (LAC), this study found that Central America and the Caribbean—the two subregions identified as the most vulnerable to oil price risk—have enormous potential to diversify away from oil-based power generation using renewable sources. In Central America, hydropower and geothermal have significant generation potential, while in the Caribbean, biomass, in the form of sugarcane bagasse, and geothermal are options with economic and technical viability. In combination with non-oil conventional alternatives, particularly natural gas, increasing the share of renewable energy sources would yield energy-security, economic, and environmental benefits. As mentioned in chapter 9, the fuel savings from a 10 percent increase in the potential capacity of renewables in Central America and the Caribbean could amount to 14.2 million and 5.6 million barrels of diesel and heavy fuel oil (HFO), respectively, representing a reduction of several points of GDP in the countries' current account. A further benefit would be a reduction in greenhouse gas (GHG) emissions.

These subregions could also reduce oil production and consumption by investing in greater energy efficiency. Various financially feasible investments are available within the power sector (supply-side efficiency) and in the transformation of electricity into energy services (demand-side efficiency). The respective savings in barrels of diesel and HFO could total 3.5 million and 1.5 million on the supply side and 9 million and 2.4 million on the demand side.

Finally, the regional integration of electricity systems via transmission lines and market agreements can reduce countries' risk exposure to the extent that they have access to more diverse, and thus less oil-dependent, regional electricity systems. As discussed in chapter 8, efforts are being made to strengthen Central America's existing regional electricity system, and electricity interconnections have been proposed for the Caribbean. The gas pipeline interconnections proposed for both subregions could help to mitigate portfolio risks, depending on the degree of correlation between natural gas and oil prices; and increased integration could lead to economies of scale that favor the use of hydropower and other renewable sources. For Central America alone, the estimated annual savings from regional electricity integration represent a reduction of about 8 percent in the oil-fired share of these countries' energy matrix.

The aggregate effect of implementing these measures would be less vulnerability to higher and more volatile oil prices, along with the development co-benefits of reduced energy expenditure by the population and climate change mitigation. While the calculations presented in this study have focused on Central America and the Caribbean, the underlying principles of the policy recommendations can be applied to any oil-importing country seeking to mitigate oil price risk. Given the far-reaching, adverse effects of high and volatile oil prices on such economies, particularly poorer or heavily indebted nations, the potential savings from implementing such measures could offer substantial macro and micro benefits, ranging from long-term financial viability of the national economy to a higher living standard for households.

That said, this optimistic outlook is not without its challenges. Making such a structural transition would entail considerable upfront costs to utilities, firms, and households; thus, supportive policies and regulations for renewable energy and energy efficiency would be required. In the case of the LAC region, regulatory, contracting, and licensing processes would need to be reformed to allow countries to implement their plans. In addition, enabling financial instruments that make these investments possible would be helpful. Furthermore, pricing reforms and technology standards would be needed to ensure that resources are not wasted. Finally, an appropriate regulatory framework and institutional strengthening would be required to facilitate trade between countries with differing regulatory policies and power-sector institutions.

Country Development Indicators

Table A.1 provides key development indicators for the 20 countries studied in the LAC subregions of Central America and the Caribbean.

Table A.2 provides key development indicators for other economies in the subregions, which were not included among the 20 countries studied.

Table A.1 Development Indicators for 20 Countries in Central America and the Caribbean

Country	Population, 2008 (thousands)	Rural population (%)	PPP GDP/capita, 2005 (US$)	Electricity capacity, 2007 (MW)	Electricity generation, 2007 (GWh/yr)
Central America					
Belize	301	48	7,325	74	213
Guatemala	13,002	51	4,185	2,140	8,425
El Salvador	7,066	39	6,367	1,419	5,560
Honduras	7,639	52	3,555	1,599	6,069
Nicaragua	5,786	43	2,410	894	3,286
Costa Rica	4,196	40	8,739	2,103	8,861
Panama	3,310	26	8,354	1,468	6,271
The Bahamas	335	16	25,784	493	2,045
Caribbean					
Jamaica	2,689	47	7,121	854	7,480
Haiti	9,780	53	1,088	244	570
Dominican Republic	9,837	31	7,595	5,518	14,840
St. Kitts and Nevis	49	68	14,939	34	195
Antigua and Barbuda	86	70	19,766	51	318
Dominica	73	26	8,033	15	84
St. Lucia	170	72	9,153	53	346
Barbados	255	60	19,397	162	950
St. Vincent and the Grenadines	109	53	8,452	23	138
Grenada	106	69	7,888	28	170
Guyanas					
Guyana	763	72	2,345	308	870
Suriname	515	25	6,938	389	1,620

Sources: World Development Indicators; OLADE 2008; Nexant 2010; EIA (www.eia.gov); www.indexmundi.com/.

Note: These 20 countries, along with Colombia, Cuba, Mexico, and Trinidad and Tobago, comprise the Association of Caribbean States (ACS); two other relevant regional organizations are (i) the Caribbean Community (CARICOM), which includes Belize, Guyana, Suriname, and all Caribbean countries, except for Cuba and the Dominican Republic and (ii) the Central American Integration System (SICA), which includes all seven Central American countries (including Belize), plus the Dominican Republic. These subdivisions reflect sets of cultural and historical ties. PPP = Purchasing Power Parity.

Table A.2 Key Development Indicators for Other Economies in the Subregions

Country or territorial unit	Population, 2008 (thousands)	Rural population, 2008 (%)	PPP GDP/capita, 2005 (US$)	Electricity capacity, 2007 (MW)	Electricity generation, 2007 (GWh/yr)
Caribbean countries					
Cuba	11,247	24	—	5,430	17,620
Trinidad and Tobago	1,338	87	22,875	1,425	6,900
Caribbean territorial units					
Bermuda (UK)	67	—	—	—	—
Turks and Caicos (UK)	—	—	—	—	—
Puerto Rico (USA)	3,954	2	—	—	23,840
Cayman Islands (UK)	54	0	—	—	—
U.S. Virgin Islands (USA)	110	5	—	—	—
British Virgin Islands (UK)	—	—	—	—	—
Anguilla (UK)	—	—	—	—	—
St. Martin (France)	—	—	—	—	—
St. Barthélemy (France)	—	—	—	—	—
Montserrat (UK)	—	—	—	—	—
Guadeloupe (France)	—	—	—	241	1,609
Martinique (France)	—	—	—	229	1,488
Aruba (Netherlands)	103	—	—	—	—
Netherlands Antilles (Netherlands)[a]	194	7	—	—	—
Guyanas territorial unit					
French Guyana	—	—	—	—	—

Sources: World Development Indicators; OLADE 2008; Nexant 2010.

Note: — = not available. PPP = Purchasing Power Parity.

a. As of October 10, 2010, the Netherlands Antilles ceased to exist; Curaçao and St. Maarten became autonomous countries, while the remaining islands (Saba, Bonaire, and St. Eustatius) were designated as special municipalities of the Netherlands.

References

Nexant. 2010. *Caribbean Regional Electricity Generation, Interconnection and Fuels Supply Strategy: Interim Report.* Prepared for the World Bank (January). San Francisco, CA: Nexant.

OLADE (Latin American Energy Organization). 2008. *Energy Statistics Report—2007.* Quito: Organización Latinoamericana de Energía.

World Bank Advisory Services: Managing Oil Price Exposure and Implementing Risk Management Strategies

Oil prices, which have been variable since the 1970s and 1980s, affect macroeconomic performance, consumers, and many other aspects of the economy. Efforts to maintain price stability can carry high costs and risks for governments. The extreme price fluctuations experienced in 2008–09 have led to increased interest in risk management strategies that can help governments limit variability in fiscal expenditures and insure against short-term price volatility. The World Bank offers advisory services to clients interested in assessing and managing commodity price risk.

Approach

Since commodity risk management has connections to fiscal risk management, public finance, agriculture and energy policies, and market development, advisory work is carried out in close collaboration with operational teams and sector specialists in Energy and Mining, Poverty Reduction and Economic Management, Financial and Private-Sector Development, Treasury, and the IMF. Advisory work is often supported by analytical work and contributions to publications as requested by project partners.

The first step in the process is a needs assessment that evaluates the current state of existing processes and procedures, suggesting areas where external expertise could build capacity and add value to implementation of risk management programs. This is usually accomplished during a one-week mission to the client country to meet with interested stakeholders. An advisory engagement is then customized to meet the specific needs and timing of the client. This can include technical support to help develop a risk management framework and implement a risk management strategy.

Needs Assessment

The objective of the needs assessment stage of the process is to assess capacity and evaluate institutional issues that can affect implementation. The specific activities involved at this stage may include:

- Analyzing the supply chain and market issues, including an overview of local, regional, and global market conditions and discussions with local and regional market participants.
- Analyzing the existing policy framework, including review of price stabilization funds, tax issues, trade regulations, and relationships between public and private actors.
- Identifying data needed to complete a comprehensive risk assessment that would quantify the specific price exposures faced by each actor in the supply chain (i.e., from government to importers to distributors to consumers).
- Defining the objectives of a hedging strategy.
- Assessing legal and regulatory infrastructure, including review of the legal/regulatory framework and/or approvals needed to support hedging transactions.
- Reviewing commodity hedging products generally, including the advantages and disadvantages of various approaches.
- Reviewing other institutional issues that would affect implementation, including governance issues related to decision-making.

The time required for the complete needs assessment is three to six weeks. This includes the desk review of sector/market issues, the one-week mission, and two to three weeks to finalize the needs assessment report.

Advisory Engagement

The objective of more in-depth advisory engagement is to provide technical support to the government for implementing a risk management strategy. Advisory services are generally fee-based, with fees paid by the recipient country and/or trust funds or other capacity-building programs. The specific activities included in this stage can include:

- Risk assessment, which carefully identifies and quantifies the government's price exposure, and how it is affected by financial actions of other actors within the supply chain.
- More detailed technical analysis to simulate the country-specific impact of price movements; examples could include:
  - comparing strategies based on buffer, earmarked funds or stabilization funds versus transferring risk to markets through commodity hedging transactions,
  - comparing short- and medium-term (one-to-seven-year), derivatives-based strategies and analyzing the effect on future expenditures, and
  - using a simulation model to measure the value of a set of generic hedging strategies under various scenarios for future oil prices.
- Assistance to revise policy frameworks, institutional frameworks, and/or support reform plans (i.e., for existing price stabilization funds).
- Development of a comprehensive commodity risk management framework that can be used on a consistent and sustainable basis. This would take into account macroeconomic and market constraints and provide a framework for assessing the cost-risk trade-offs necessary to guide decisionmaking.
- Education for stakeholders and policy makers.
- Technical capacity-building for ministry staff responsible for implementing the strategy and execution of transactions; this could include:
  - guidance on managing operational risk;
  - advice on assessing market liquidity, handling relationships with banks, and evaluating pricing; and
  - assessing the trading platform for financial derivatives (front-office, back-office, accounting, and legal), if applicable.
- Technical support to design a framework and governance process for selecting hedging strategies; this could include:
  - evaluating alternatives based on minimizing the present value of cash flows generated by oil expenditures and hedging instruments

under various price scenarios over a short-to-medium time horizon, subject to the client's risk tolerance; and

o assistance in evaluating instruments for implementing derivatives-based strategies (futures, OTC forward or options, swaps, and structured products).

- Technical support for any required changes in legal and regulatory infrastructure.
- Technical advice related to structuring and executing transactions with the market.
- Market execution on behalf of the client.

The time required for the advisory engagement, comprising a combination of field work and long-distance support, depends on the scope and customization of client needs.

Accounting for Fuel Price Risk in Power System Planning: Case Studies in the Dominican Republic and Honduras

Portfolio analysis is a powerful way to identify the systematic risk of fuel prices to power-system investments, based on analysis of a country's choices for expanding its power system, primarily using the Wien Automatic System Planning (WASP). In addition, portfolio optimization can help to identify feasible generation portfolios superior to those chosen by the WASP or system planners, through utilizing tools highly accessible to system planners, financial analysts, and others.

This appendix presents the results of a portfolio optimization exercise designed to account for fuel price risk in power system planning. Implemented in the Dominican Republic and Honduras, the exercise considers a series of oil price scenarios to quantify gains from a more diversified electricity generation matrix, as described in chapter 6.[1] Applying the approach yields explicit trade-offs between risk and return, as a supplement to least-cost optimization models (primarily the WASP).[2] The findings confirm the benefits of diversifying power generation, highlighting the feasible options that countries as distinct as the Dominican Republic and Honduras have available (box C.1).

Box C.1

Interpreting Portfolio Risk Results

The model calculates portfolio risk as the square root of the total variation in a given portfolio, according to the historical database on each risk factor. For the Dominican Republic in 2010, for example, US$12.6 million represents the *expected risk* of the least-cost portfolio of generation units. Other portfolios for 2010 would show lower risk exposure, *if feasible*. The total exposure of the system to fuel price risk is not only the expected risk, but also the calculated cost differential for that portfolio at higher fuel prices. For the Dominican Republic's 2010 base case, simply increasing the price of crude oil to US$90 per bbl would raise the cost of the existing generation portfolio by $330 million, for a total of $2.15 billion. This added cost, all of it for fuel, represents the calculated fuel price exposure, and is distinct from the expected risk. Therefore, it is important to use realistic oil-pricing scenarios to calculate efficient generation portfolios and thus capture risk exposure correctly.

Source: Hertzmark 2010.

Methodology

This exercise evaluated the impacts of various technologies and fuel price strategies on the (i) overall cost of supplying power and (ii) country exposure to fuel price swings resulting from choices in generation options. For the Dominican Republic and Honduras, base cases were constructed for each country using 2008–09 data; the 2010 base cases matched the estimated capacity and energy demand for base, intermediate, and peak loads, at a price of US$75 per barrel.

The Dominican Republic: System Overview and Simulation Results
In the Dominican Republic, system peak demand was 5.3 GW in early 2010. Electricity-generation plants rely primarily on fossil fuels. Oil and gas, operated in baseload mode, are used to meet more than half of all electricity demand. Steam-turbine baseload plants using heavy fuel oil (HFO) have the largest share of any type of plant, at 28.7 percent; while coal meets 14 percent of demand operating in the baseload model. Hydropower, mainly in cycling mode, meets 17 percent of capacity needs used for the equilibrium solution (P_{oil} = US$75 per bbl) (table C.1).

Table C.1   Share of Plant Types to Meet Energy Demand

Power plants and imports	Share (% energy)
HFO units	28.7
Gas turbines (diesel and gas)	9.5
Coal steam	14.1
Combined-cycle units	24.5
Hydropower	17.1
Other (wind, municipal solid waste)	1.9

Source: CNE-DR, Electricity System Expansion Plan (*Prospectiva 2009*).

The base case shows the kilowatt hours generated to meet base, intermediate, peak, and total demand for both capacity and energy. The capacities for the various generation technologies are limited to those currently in place, with constraints on both hydropower (i.e., water availability) and gas (i.e., LNG supply).[3] Since all plants are existing, fixed charges are limited to fixed operations and maintenance (O&M); while variable costs cover fuel, variable O&M, and spinning reserve margin.

Demand growth rates for electricity, provided by the Electricity System Expansion Plan,[4] average 4.25 percent annually through 2020. Assumptions in the 2010 base case conform to those in *Prospectiva 2009* with regard to water availability (50 percent of maximum plant capacity, on average), coal restrictions (none), gas availability (non-binding), and plant retirements (not applicable). For subsequent simulation years, these conditions are generally maintained as base-case conditions for 2015 and 2020.[5]

Plant characteristics are identical to those used in *Prospectiva 2009* with regard to combustion efficiency, duty cycle, and lifetime. However, the costs provided in *Prospectiva 2009* were considered low by current standards, leading to potential errors in choosing efficient portfolios. Consequently, plant investment costs, expressed as US dollars per kilowatt, were adjusted upward to reflect current cost guidelines for each plant type.

With little land area and no possibility of importing electricity, the Dominican Republic's electric power system is constrained. Thus, generation solutions requiring significant additional land resources (e.g., onshore wind farms, coal import and storage, LNG re-gasification, oil storage, hydro with dam storage [including pumped hydro]) may likewise be limited.

Physical limitations on generation options. In the exercise, such technologies as wind generation, standard coal-steam turbine fuel cycles, and excess LNG re-gasification capacity were limited with regard to ultimate deployment potential. The country's power system, with its relatively high demand and space limitations, depends on fuel-supply logistics for access to generation resources that are more attractive than HFO. This means that LNG re-gasification, as well as coal import and storage, capacity are vital to providing the country some degree of diversification in energy resources.

Fuel prices. Fuel prices were drawn from the 2008 ESMAP Caribbean database, updated to 2010 price levels. Prices for LNG are based on U.S. Gulf Coast netbacks. For all of the model's base-case runs, it was assumed that hydrocarbon fuel prices would vary with their historic characteristics, with the caveat that two circumstances could alter the pricing risk for certain fuels: (i) shale-gas production, which could induce the flattening of the pricing curve for LNG, reducing its volatility over a large range and (ii) hedging contracts, which, at a certain price, would similarly dampen the volatility of a specific fuel.

Risks and costs of electricity supply, 2015 and 2020. The model, run for the base-case set of parameters for 2015 and 2020, provides a snapshot of least-cost and -risk investment options for each of those years. Generally, the 2020 results are additive to the 2015 results, given the basic assumption that new plants cannot be "unbuilt" to meet changing price or risk conditions. For 2015, the system must be built out to meet a demand of 8.3 GW, 3.8 GW more than in 2010. Table C.2 shows the range of results from running oil price simulations at US$55–110 per barrel.

These results suggest that riskier portfolios are optimal if one believes the price of oil will remain in the range of US$55–65 per barrel. In those cases, gas-fired, combined-cycle gas turbine (CCGT) is the most cost-

Table C.2   Results of Base-Case Simulations: Dominican Republic, 2015

Variable	Cost				
Oil price (US$/bbl)	55	65	75	90	110
Total generation cost (US$, billions)	1.85	2.12	2.34	2.64	2.99
Total portfolio risk (US$, millions)	14.6	13.7	12.7	12.6	12.6
Average generation cost (¢/kWh)	5.0	5.7	6.4	7.0	8.1
Average new generation cost (¢/kWh)	5.8	6.4	6.9	7.4	7.9

Source: Hertzmark 2010.

effective generation option. Only at the lowest oil price is HFO operated to any significant extent. In all cases at or above US$65 per barrel, HFO plants are generally phased out.

For all simulations, cost does not rise as fast as the price of oil because, as the cost of oil rises, the cost-minimizing portfolio moves toward the risk-minimizing portfolio (i.e., more hydro and solid fuels). Thus, the total portfolio risk at higher oil prices is lower, given less exposure to oil price fluctuations. At the two highest oil prices, US$90 and $110 per barrel, the model uses all of the coal and hydro available by that date. One should also note the shrinking gap between total supply cost and new plant cost throughout the scenarios. When oil is inexpensive, new generation costs are dominated by the cost of the investment itself. Existing power plants are the least expensive units to operate. As the price of fuel climbs, the gap between old and new plants, almost 1¢ per kWh at first, shrinks to less than 0.25¢ per kWh and then reverses at the highest oil price, as new plants, with their greater efficiencies, prove less susceptible to fuel-price woes than the current fleet of plants.

Coal and hydro minimize both cost and risk under scenarios at or above US$75 per barrel. Initially, the country's coal technology would be steam turbine, featuring improved efficiency compared with the current steam-boiler combustion using coal.[6] After 2015, investment in coal would shift increasingly to integrated gasification combined cycle (IGCC), which would offer not only reduced emissions but also greatly improved throughput efficiency.[7]

Hydro development remains limited simply by resource and land availability. Post-optimality tests on various scenarios indicate that the value of additional hydro capacity to reduce both cost and risk is among the highest of all available resources.

For the 2020 scenarios, additional technologies, including IGCC, are more widely available. In addition, there is a full build-out of the country's remaining hydro potential. Combustion turbine units are limited to peak-period duty cycles and emergency backup (table C.3).

For 2020, the country's system is constrained by the ability to construct new plants not based on coal or CCGT. At the low end of oil price simulations, CCGT plants are built at a rate that more than doubles existing LNG imports. At the high end, coal-fuel cycles account for at least half of total capacity. With the move toward coal-fired generation, the cost of supply (both total and new kilowatt hours) does not rise at nearly the same pace as that of oil. An appropriate forecast for oil prices is

Table C.3   Results of Base-Case Simulations: Dominican Republic, 2020

Variable	Cost				
Oil price (US$/bbl)	55	65	75	90	110
Total generation cost (US$, billions)	2.44	2.77	3.01	3.35	4.01
Total portfolio risk (US$, millions)	16.6	16.0	14.5	14.3	12.9
Average generation cost (¢/kWh)	5.2	5.9	6.4	7.1	7.5
Average new generation cost (¢/kWh)	5.9	6.5	7.0	7.5	7.5

Source: Hertzmark 2010.

critical for the country. Without making a major commitment to either LNG or coal over the next several years, the country runs the risk of not meeting demand in 2020. Without enough new, efficient generating resources, retirements of existing gas-turbine and HFO units would need to be delayed, greatly increasing generation cost and exposure to fuel-price fluctuations. Scenarios were run with limitations on both gas and coal plant construction. Hydro capacity is fully built out in all scenarios.

Wind contributes modestly in 2020, at 425 MW, and is constrained by limited land availability in all scenarios.[8] Without significant new hydro-capacity additions after 2015, the mirroring of wind falls to gas turbines and diesel engines, significantly increasing the cost of using wind energy.

Honduras: System Overview and Simulation Results

Compared to the Dominican Republic, Honduras faces fewer absolute limitations on making choices about its future power system. It has import options from the Central American Power Pool (CAPP), untapped hydro-power and geothermal potential, and a smaller legacy generation fleet.[9]

According to the regional indicative plan, the country's current peak demand of 1.47 GW is projected to rise to 2.01 GW and 2.76 GW by 2015 and 2020, respectively. Natural gas is expected to play a role in the generation supply mix once a Central American re-gasification plant, to be constructed along the Caribbean coast, enters into operation by mid-decade.

Risks and costs of electricity supply, 2015 and 2020. The model was run for the base-case set of parameters for 2015 and 2020, meaning that the model provides a snapshot of least-cost and -risk investment options for those respective years. Like the Dominican Republic case, the 2020 results are additive to the 2015 results, given the basic assumption that new plants cannot be "unbuilt" to meet changing price or risk conditions. For 2015, the system must be built out to meet a demand of 2.76 GW,

Table C.4   Results of Base-Case Simulations: Honduras, 2015

Variable	Cost				
Oil price (US$/bbl)	55	65	75	90	110
Total generation cost (US$, millions)	464	512	543	575	604
Total portfolio risk (US$, millions)	2.45	2.35	1.76	1.75	1.75
Average generation cost (¢/kWh)	4.6	5.1	5.1	5.4	5.7
Average new generation cost (¢/kWh)	5.5	5.9	6.3	6.6	6.9

Source: Hertzmark 2010.
Note: Includes 200 MW of negawatts.

1.3 GW more than in 2010. Table C.4 shows the range of results from running oil price simulations at US$55–110 per barrel.

These results show that the riskier portfolios are optimal if one believes the price of oil will remain within the range of US$55–65 per barrel. In those cases, gas-fired CCGT is a cost-effective generation option, as are existing HFO plants; however, coal is not preferred under low-price conditions. Hydro is always chosen at its maximum level; while geothermal, the Northern Interconnect from CAPP, and negawatts (capacity savings due to conservation measures) are all chosen as system resources under all oil price scenarios. In all cases at or above US$75 per barrel, HFO plants are generally phased out.

Negawatts cut demand for intermediate and peak loads by about 10 percent; without this system resource, both cost and risk rise. The negawatt program was assumed to cost US$2,500 per kW saved, a conservative assumption relative to some claims of negative costs for conservation. In the 2015 scenarios with oil at US$110 per barrel, 200 MW of negawatts reduces total cost by nearly 10 percent and risk by about 15 percent. Without the negawatt program, the average cost of generation would rise from 5.7¢ to 6.04¢ per kWh.

Like the Dominican Republic case, oil cost does not rise as fast as price for all simulations. As cost increases, the cost-minimizing portfolio moves toward the risk-minimizing portfolio, meaning that, at higher oil prices, total portfolio risk is lower with less exposure to oil price fluctuations. At the two highest oil prices, US$90 and $110 per barrel, the model uses all of the coal and hydro available by that date. Unlike the Dominican Republic case, the gap between average and new generation costs does not narrow as the price of oil rises because most of the low/no-fuel cost options are used, even when oil is at US$55 per barrel.

Coal and hydro minimize both cost and risk under scenarios at or above US$75 per barrel. Unlike the Dominican Republic case, the coal technology of choice in Honduras is primarily IGCC, which, compared to current steam-boiler combustion using coal, features better efficiency and significantly lower emissions.

Hydropower development is aggressive in the base case, providing far lower average generation costs than in the Dominican Republic case. Hydro, negawatts, and geothermal remain the most cost-effective system resources for both the 2015 and 2020 simulations.

For the 2020 scenarios, additional technologies, including more negawatts, geothermal, and IGCC, are available. There is also potential to build out the country's remaining hydro potential. Combustion turbine units are limited to peak-period duty cycles and emergency backup (table C.5).

Honduras exhibits no serious constraints in 2020. Even when oil reaches US$110 per barrel, the hydro potential is not fully built out, owing primarily to the energy and capacity from negawatts and the regional interconnection. If the negawatts (400 MW of intermediate and peak demand equivalent in 2020) were excluded, hydropower, as well as IGCC and coal, would have to be fully built out. Some wind would be used (2 percent of total capacity), especially since hydro acts as the mirroring resource in Honduras.

For the 2020 simulation of US$90 per barrel without negawatts, the total annual supply cost would rise by US$30 million and risk would increase by 5 percent. The unit cost of generation would rise by 0.2¢ per kWh on average, and generation from new sources would remain constant, owing, in large part, to the hydro-capacity build-out.

Moderating costs and risks. Unlike the Dominican Republic, Honduras does not face high costs and risks or having to make irreversible decisions. Rather, the country's major challenge is constructing hydropower plants

Table C.5   Results of Base-Case Simulations: Honduras, 2020

Variable	Cost				
Oil price (US$/bbl)	55	65	75	90	110
Total generation cost (US$, millions)	663	726	760	809	858
Total portfolio risk (US$, millions)	3.29	3.22	2.53	2.53	2.23
Average generation cost (¢/kWh)	4.8	5.2	5.3	5.6	5.8
Average new generation cost (¢/kWh)	5.4	5.8	6.1	6.5	6.7

Source: Hertzmark 2010.
Note: Includes 400 MW of negawatts

Table C.6   Results of Hydro Generation and Negawatt Limitations, 2020

Measure of merit	US$65 per bbl	US$90 per bbl
Total generation cost (US$, millions)	794	898
Total portfolio risk (US$, millions)	4.10	3.15
Average generation cost (¢/kWh)	5.7	6.1
Average new generation cost (¢/kWh)	6.4	7.0

Source: Hertzmark 2010.

on schedule. That is, if new hydro-plant capacity in 2020 were limited to 375 MW, instead of the 697 MW used in the base-case scenario, generation costs would rise sharply, requiring more HFO and gas (table C.6).

With less hydro and conservation, coal fills the supply gap, with 35 percent of generation capacity. At US$90 per barrel, HFO is not an attractive generation resource. In 2020, the overall supply cost rises by 10 percent (nearly US$90 million). At lower oil prices, the reduced availability of hydro and negawatts would lead to greater use of oil and gas and thus create more risk in the generation portfolio.

In the simulation at US$65 per barrel, coal would fall back to its more typical proportion at low oil prices, representing 14 percent of total capacity. CCGT would rise to 14 percent, while existing HFO plants would comprise 20 percent of total supply. Compared to the unconstrained base case at US$65 per barrel, the total supply cost would rise by US$59 million (8.1 percent) and generation unit costs would rise by nearly 10 percent on average (for new generation supplies, the cost increase would exceed 10 percent).

A worst-case scenario for Honduras would feature limited hydro development, no negawatts, and a moratorium on new coal plants (even IGCCs). With oil at US$90 per barrel, the system would still rely heavily on older HFO plants for 20 percent of total capacity and CCGTs for 19 percent of capacity; wind would be expanded to the maximum, at 7 percent of total capacity, and costs and risks would rise significantly relative to the base-case scenario (table C.7).

Even at oil prices considered moderate by current standards, limited access to hydro and coal generation would increase supply costs considerably. Compared to the base case, these limitations would increase total generation costs by 12 percent; generation unit costs would rise by 8.8 percent, on average, and by 13.7 percent for new capacity. The riskiness of the portfolio would expand by about half, compared with the base case.

Table C.7   Results of Hydro and Coal Generation and Negawatt Limitations, 2020

Measure of merit	US$65 per bbl	US$90 per bbl
Total generation cost (US$, millions)	812	1,010
Total portfolio risk (US$, millions)	4.66	4.66
Average generation cost (¢/kWh)	5.9	7.3
Average new generation cost (¢/kWh)	6.6	7.9

Source: Hertzmark 2010.

At US$90 per barrel, the additional costs and risks for the Honduras system would be extremely high. Without a cost-and-risk minimizing option for coal and hydro, the generation system would be essentially the same as with oil at US$65 per barrel. In this case, the total annual supply cost would rise from US$809 million in the base case to US$1.01 billion in the restricted case, representing a 25 percent increase. Unit generation costs would increase by 30 percent on average, and by 18 percent for new generation sources. The riskiness of the overall portfolio, US$2.53 million in the base case, would rise by more than three-quarters, indicating the system's high exposure to fuel price fluctuations.

Among the three factors that moderate risk for Honduras—hydropower, negawatts, and coal—coal is the most significant quantitatively. Its absence would account for roughly 55 percent of total increased costs and most of the increased fuel price risk relative to hydropower and conservation.

Notes

1. The simulation results presented in this appendix are excerpted from Donald Hertzmark, "Accounting for Fuel Price Risk in Electric Power System Planning: Case Studies of Honduras and the Dominican Republic," Energy Sector Management Assistance Program (ESMAP), World Bank (June 2010).

2. The analysis included portfolio analysis (the study of the risk and return of alternative WASP generation portfolios) and portfolio optimization (the construction of optimal risk-return portfolios that can be further investigated using WASP or similar tools).

3. The gas-supply constraint is expressed in the model as the total kilowatt hours generated from gas. The model allocates existing gas kilowatt hours to their most efficient generation resources. The underlying gas supply constraint is expressed in MMBtu of LNG and is converted into kilowatt hours for purposes of the model. This method permits the model to work only in

its appropriate energy units and avoids dimensional difficulties with the Excel Solver program

4. *Prospectiva 2009*.

5. That is, the construction of coal-fired power plants would not be limited by external factors, older gas and HFO steam units would be retired (as per *Prospectiva 2009*), and the duty cycles of remaining plants of that type would be modified as appropriate (e.g., older HFO plants would shift to cycling use rather than baseload, where possible).

6. Super critical (efficiency of 42 percent) and ultra-super critical (efficiency of 44–45 percent).

7. IGCC efficiencies are 47–51 percent for the full cycle, including gasification.

8. Additional wind scenarios were run at various levels of mirroring. If mirroring of wind is limited to 50 percent of output, the system can absorb as much as 700 MW of wind by 2020, instead of the 425 MW in the base case. Generation costs per kilowatt hour will rise slightly in the case of US$90 per barrel. If wind mirroring rises to 75 percent, total generation costs will increase relative to the country base case.

9. In the Honduras case study, we used recent U.S. EIA data on the country's electricity demand and generation and national studies on demand and power-plant costs; we considered a median demand growth rate of 4.9 percent.

Reference

Hertzmark, D. 2010. *Accounting for Fuel Price Risk in Electric Power System Planning: Case Studies of Honduras and the Dominican Republic*. Report prepared for the Energy Sector Management Assistance Program (ESMAP). Washington, DC: World Bank.

Bibliography

Aburto, J. 2010. *Study on Mitigating Impact of Oil Price Volatility in Central America and the Caribbean Energy Efficiency of Public Utilities: Transmission and Distribution Losses*. Report prepared for the World Bank, Washington, DC: World Bank.

Anderson, P. R. D., A. C. Silva, and A. Velandia-Rubiano. 2010. "Public Debt Management in Emerging Market Economies: Has This Time Been Different?" World Bank Policy Research Working Paper 5399. Washington, DC: World Bank.

Arezki, R., and M. Bruckner. 2010. "International Commodity Price Shocks, Democracy, and External Debt." IMF Working Paper WP/10/53, International Monetary Fund, Washington, DC.

Awerbuch, S. 2000. "Getting It Right: The Real Cost Impacts of a Renewables Portfolio Standard." *Public Utilities Fortnightly*, February 15.

Awerbuch, S., and M. Berger. 2003. "Applying Portfolio Theory to EU Electricity Planning and Policy Making." IEA/EET Working Paper, International Energy Agency, Paris.

Bacon, R., and M. Kojima. 2006. *Coping with Higher Oil Prices*. Energy Sector Management Assistance Program (ESMAP) Report 323/06. Washington, DC: World Bank.

————. 2008a. *Coping with Oil Price Volatility*. Energy Sector Management Assistance Program (ESMAP) Energy Security Special Report 005/08. Washington, DC: World Bank.

————. 2008b. *Oil Price Risks: Measuring the Vulnerability of Oil Importers*. Public Policy for the Private Sector, Note No. 320. Washington, DC: World Bank Group.

————. 2008c. *Vulnerability to Oil Price Increases: A Decomposition Analysis of 161 Countries*. Extractive Industries and Development Series #1. Washington, DC: World Bank Group.

Bacon, R., and A. Mattar. 2005. *The Vulnerability of African Countries to Oil Price Shocks: Major Factors and Policy Options*. Energy Sector Management Assistance Program (ESMAP) Report 308/05. Washington, DC: World Bank.

Barsky, R., and L. Kilian. 2004. "Oil and the Macroeconomy Since the 1970s." *Journal of Economic Perspectives* 18(4): 115–34.

Batini, N., and E. Tereanu. 2009. "What Should Inflation Targeting Countries Do When Oil Prices Rise and Drop Fast?" IMF Working Paper WP/09/101, International Monetary Fund, Washington, DC.

Bazilian, M., and F. Roques, eds. 2008. *Analytical Methods for Energy Diversity and Security. Portfolio Optimization in the Energy Sector: A Tribute to the Work of Dr. Shimon Awerbuch*. Amsterdam: Elsevier.

Bernanke, B., M. Gertler, M. Watson, C. Sims, and B. Friedman. 1997. "Systematic Monetary Policy and the Effects of Oil Price Shocks." *Brookings Papers on Economic Activity* 1997(1): 91–157.

Blanchard, O., and J. Gali. 2007. "The Macroeconomic Effects of Oil Price Shocks: Why Are the 2000s So Different from the 1970s?" In *International Dimensions of Monetary Policy*, ed. J. Gali and M. J. Gertler, 373–421. Cambridge, MA: National Bureau of Economic Research.

Cavallo, M. 2008. *Oil Prices and Inflation*. FRBSF Economic Letter No. 2008-31. San Francisco, CA: Federal Reserve Bank of San Francisco.

CEAC (Central American Electrification Council). 2007. "Plano Indicativo Regional de Expansion de la Generación Periodo 2007–2020." Grupo de Trabajo de Planificación Indicativa Regional (GTPIR), Consejo de Electrificación de America Central, Tegucigalpa.

Chen, S. S., and H. C. Chen. 2007. "Oil Prices and Real Exchange Rates." *Energy Economics* 29(3): 390–404.

CNE-DR (National Energy Commission-Dominican Republic). 2006. *Plan Indicativo de la Generación del Sector Eléctrico Dominicano*. 2006–20 study period, 2005 annual adjustment. Santo Domingo: Comisión Nacional de Energía de la República Dominicana.

CONACE (National Commission on Energy Conservation). 2003. *Programa Nacional de Conservación de Energía 2003–2008 (PRONACE)*. San José: Comisión Nacional de Conservación de Energía.

Consejo de Planificación de Electrificación de América Central. 2009. "Plan Indicativo Regional de Expansión de la Generación, 2009–2023," May.

Cuevas, F., V. H. Ventura, E. Rojas, and J. Alvarado. 2009. *Istmo Centroamericano: Las Fuentes Renovables de Energía y el Cumplimiento de la Estrategia 2020.* ECLAC Report LC/MEX/L.953. Mexico City: UN Economic Commission for Latin America and the Caribbean.

Dana, J. 2010. *Hedging Commodity Price Volatility with Market-Based Tools.* World Bank internal report. Washington, DC: World Bank.

ECLAC (UN Economic Commission for Latin America and the Caribbean). 2009a. *La Crisis de los Precios del Petróleo y su Impacto en los Países Centroamericanos.* Report No. LC/MEX/L.908, June 18. Mexico City: UN Economic Commission for Latin America and the Caribbean.

———. 2009b. "Situación y Perspectivas de la Eficiencia Energética en América Latina y El Caribe." Prepared for the Regional Intergovernment Meeting on "Energy Efficiency in Latin America and the Caribbean", Santiago, Chile. September 15–16.

ECLAC (UN Economic Commission for Latin America and the Caribbean) and SICA (Central American Integration System). 2007. "Estrategia Energética Sustentable Centroamericana 2020." UN Economic Commission for Latin America and the Caribbean and General Secretariat of the Central American Integration System, Mexico City.

Econoler. 2010. *Market Assessment for Promoting Energy Efficiency and Renewable Energy Investment in Brazil through Local Financial Institutions.* Report prepared for the International Finance Corporation. Quebec City: Econoler.

Economist. 2010a. "This Changes Everything," 16, North American Edition, March 13.

———. 2010b. "An Unconventional Glut," 72–74, North American Edition, March 13.

EIA (U.S. Energy Information Administration). 2006. Statistical Database. http://www.eia.gov.

Eletrobrás. 2009. *Resultados do PROCEL—2008.* Eletrobrás: Rio de Janeiro.

ESMAP (Energy Sector Management Assistance Program). 2007. "Technical and Economic Assessment of Off-grid, Mini-grid, and Grid Electrification Technologies." Energy Sector Management Assistance Program, Technical Paper 121/07, World Bank, Washington, DC.

————. 2008. *Risk Assessment Methods for Public Utilities.* Energy Sector Management Assistance Program. Washington, DC: World Bank.

Ferderer, P. 1996. "Oil Price Volatility and the Macroeconomy." *Journal of Macroeconomics* 18(1): 1–26.

Fukunaga, I., N. Hirakata, and N. Sudo. 2010. "The Effects of Oil Price Changes on the Industry-level Production and Prices in the U.S. and Japan." NBER Working Paper 15791, National Bureau of Economic Research, Cambridge, MA.

Gerner, F. 2010. *Caribbean Regional Electricity Generation, Interconnection and Fuels Supply Strategy: Synthesis Report.* Prepared with support of the LCR Energy Group. Washington, DC: World Bank.

Gerner, F., and M. Hansen. 2011. *Caribbean Regional Electricity Supply Options: Toward Greater Security, Renewables, and Resilience.* Report No. 59459. Washington, DC: World Bank.

Gilbert, C. 2010. "Electricity Hedging." Background research prepared for the study, Mitigating Vulnerability to High and Volatile Oil Prices in Latin America and the Caribbean, World Bank, Washington, DC.

GPL (Guyana Power and Light, Inc.). 2006. *Annual Report.* Georgetown: Guyana Power and Light.

Hamilton, J., and A. Herrera. 2004. "Oil Shocks and Aggregate Macroeconomic Behavior: The Role of Monetary Policy." *Journal of Money, Credit and Banking* 36(2): 265–86.

Herbst, A. 1990. *The Handbook of Capital Investing.* New York: Harper Business.

Hertzmark, D. 2010. *Accounting for Fuel Price Risk in Electric Power System Planning: Case Studies of Honduras and the Dominican Republic.* Report prepared for the Energy Sector Management Assistance Program (ESMAP). Washington, DC: World Bank.

Hunt, B., P. Isard, and D. Laxton. 2002. "The Macroeconomic Effects of Higher Oil Prices." *National Institute Economic Review* 179(1): 87–103.

IAEA (International Atomic Energy Agency). 2006. "Wien Automatic System Planning (WASP) Package: A Computer Code for Power Generation System Expansion Planning," Version WASP-IV, International Atomic Energy Agency, Vienna.

IBRD (International Bank for Reconstruction and Development). 2007. *Honduras: Power Sector Issues and Options.* International Bank for Reconstruction and Development, Washington, DC.

ICE (Costa Rican Electricity Institute). 2007. *Plan de Expansión de la Generación Eléctrica Período 2008–2021.* San José: Instituto Costarricense de Electricidad, Centro Nacional de Planificación Eléctrica.

IEA (International Energy Agency). 2004. *Analysis of the Impact of High Oil Prices on the Global Economy*. Paris: International Energy Agency.

IMF (International Monetary Fund). 2000. "Impact of Higher Oil Prices on the Global Economy." Research Department Staff Paper, International Monetary Fund, Washington, DC.

————. 2006. *Regional Economic Outlook: Sub-Saharan Africa*. World Economic and Financial Surveys. Washington, DC: International Monetary Fund.

Johnson, T. M., C. Alatorre, Z. Romo, and F. Liu. 2009. *Low-carbon Development for Mexico*. Washington, DC: World Bank.

Kojima, M. 2009. *Government Responses to Oil Price Volatility: Experience of 49 Developing Countries*. Extractive Industries for Development Series #10. Washington, DC: World Bank.

Larson, E. D., R. H. William, and M. R. Leal. 2001. *A Review of Biomass Integrated-Gasifier/Gas-turbine Combined Cycle Technology and Its Application in Sugarcane Industries, with an Analysis for Cuba*. Center for Energy and Environmental Studies. Piracicaba, Brazil: Princeton University/Centro de Tecnología Copersucar.

LAWEA (Latin America Wind Energy Association). 2010. *Energía Eólica en América Latina 2009–2010*. Guadalajara: Latin America Wind Energy Association.

Lee, K., and S. Ni. 2002. "On the Dynamic Effects of Oil Price Shocks: A Study Using Industry Level Data." *Journal of Monetary Economics* 49(4): 823–52.

Markowitz, H. 1952. "Portfolio Selection." *Journal of Finance* 7: 77–91

MEM (Ministry of Energy and Mines)-Jamaica. 2009. *Jamaica's National Energy Policy 2009–2030*. Kingston: Ministry of Energy and Mines.

NAS/NAE (National Academy of Sciences/National Academy of Engineering). 2009. *Electricity from Renewable Resources: Status, Prospects, and Impediments*. America's Energy Future Panel on Electricity from Renewable Resources, National Research Council. Washington, DC: National Academy of Sciences/National Academy of Engineering.

NEPCO (National Electric Power Company). 2007a. *Study of Alternatives Available To Meet the Demand for Primary Energy and Choose the Best Alternative*. Amman, Jordan: National Electric Power Company.

————. 2007b. *Annual Report*. Amman, Jordan: National Electric Power Company.

Nexant. 2010a. *Promoting Sustainable Energy Integration in Central America*. Assessment for USAID El Salvador and USAID Central America and Mexico Regional Program. San Francisco, CA: Nexant.

————. 2010b. *Caribbean Regional Electricity Generation, Interconnection and Fuels Supply Strategy: Interim Report*. Prepared for the World Bank (January). San Francisco, CA: Nexant.

NREL (U.S. National Renewable Energy Laboratory). 2001. *Wind Energy Resource Atlas of the Dominican Republic.* Report NREL/TP-500-27602. Golden, CO: U.S. National Renewable Energy Laboratory.

OLADE (Latin American Energy Organization). 2005. *Prospectiva Energética de América Latina y el Caribe.* Quito: Organización Latinoamericana de Energía.

————. 2008. *Energy Statistics Report—2007.* Quito: Organización Latinoamericana de Energía.

————. 2009. *Asistencia Técnica para Reducción de Pérdidas en Redes de Distribución de Nicaragua.* Quito: Organización Latinoamericana de Energía.

OUR (Office of Utilities Regulaton)-Jamaica. 2009. *Annual Report and Financial Statements 2007–08.* Kingston: Office of Utilities Regulation.

Poole, A. D. 2009. "The Potential of Renewable Energy Resources for Electricity Generation in Latin America." Working paper prepared for the report, Latin America and Caribbean Region's Electricity Challenge, World Bank, Washington, DC.

Rodas, O. R. 2008. "Avances en Uso Racional y Eficiencia Energética en Honduras." Secretaria de Recursos Naturales y Ambiente/Dirección General de Energía, II Seminario Internacional de Ahorro de Energía, PESIC, Tegucigalpa, April 17–18.

Rogozinski, J. 2000. *A Brief History of the Caribbean.* New York: Plume/Penguin Putnam.

Sauter R., and S. Awerbuch. 2003. "Oil Price Volatility and Economic Activity: A Survey and Literature Review." IEA Research Paper, International Energy Agency, Paris.

Schwerin, A. 2010. "Analysis of the Potential Solar Energy Market in the Caribbean." Caribbean Renewable Energy Development Program (CREDP)/ GTZ, July.

Sharpe, W. 2000. *Portfolio Theory and Capital Markets.* New York: McGraw-Hill.

Sharpe, W., G. Alexander, and J. Bailey. 1999. *Investments.* New Delhi: Prentice Hall.

Spatafora, N., and A. Warner. 1995. "Macroeconomic Effects of Terms of Trade Shocks: The Case of Oil-exporting Countries." Policy Research Working Paper 1410, World Bank, Washington, DC.

UN (United Nations). 2009. *Energy Balances and Electricity Profiles 2009.* New York: United Nations, Department of Economic and Social Affairs.

UNFCCC (United Nations Framework Convention on Climate Change)-Dominican Republic. 2004. *Primera Comunicación Nacional: Convención Marco de Naciones Unidas sobre Cambio Climático.* Santo Domingo: Secretaría de Estado de Medio Ambiente y Recursos Naturales.

USAID (United States Agency for International Development)-Dominican Republic. 2004. *Estrategia de Eficiencia Energética para la República Dominicana.* Santo Domingo: United States Agency for International Development and National Energy Commission.

Villafuerte, M., and P. Lopez-Murphy. 2010. "Fiscal Policy in Oil Producing Countries during the Recent Oil Price Cycle." IMF Working Paper WP/10/28. International Monetary Fund, Washington, DC.

WEC (World Energy Council). 2008. *Regional Energy Integration in Latin America and the Caribbean.* London: World Energy Council.

World Bank. 2006. *Assessing the Impact of Higher Oil Prices in Latin America.* Economic Policy Sector. Washington, DC: World Bank.

———. 2008. "GEF Project Brief on a Proposed Grant from the GEF Trust Fund to the HKJ for a Promotion of a Wind Power Market Project." World Bank, Washington, DC.

———. 2011. *Regional Power Integration: Structural and Regulatory Challenges.* Central America Regional Programmatic Study for the Energy Sector. Report No. 58934-LAC. Washington, DC: World Bank.

World Bank/ESMAP (Energy Sector Management Assistance Program). 2009. *Central America Sector Overview: Regional Programmatic Study for the Energy Sector—General Issues and Options Module.* Washington, DC: World Bank.

———. 2011. *Drilling Down on Geothermal Potential: A Roadmap for Central America.* Washington, DC: World Bank.

Yépez-García, A. 2009. *The Impact of Oil Prices on Electricity Costs in Latin America and the Caribbean.* Energy Sector LCR. Washington, DC: World Bank.

Yépez-García, R. A., T. M. Johnson, and L. A. Andrés. 2011. *Meeting the Balance of Electricity Supply and Demand in Latin America and the Caribbean.* Directions in Development. Washington, DC: World Bank.